Companion to Angular Momentum

Companion to Angular Momentum

by

Valeria D. Kleiman
National Institute of Standards and Technology
Gaithersburg, Maryland

Hongkun Park
Lawrence Berkeley National Laboratory
Berkeley, California

Robert J. Gordon
Department of Chemistry
University of Illinois at Chicago
Chicago, Illinois

Richard N. Zare
Department of Chemistry
Stanford University
Stanford, California

A Wiley-Interscience Publication
JOHN WILEY & SONS, INC.
New York/Chichester/Weinheim/Brisbane/Singapore/Toronto

ISBN 0-471-19249-X

10 9 8 7 6 5 4 3 2 1

Contents

PREFACE

For as long as angular momentum theory has been taught, students have found this topic difficult to learn and apply. Teaching experience shows that mastery of this topic comes chiefly from working through numerous examples and exercises. In many ways, learning angular momentum theory is like learning to drive a car. Formal instruction can be quite helpful, but real driving experience is essential. For that reason, Zare's *Angular Momentum* contains a large number of problems in the form of what are called "problem sets" and "applications" at the end of each chapter. This book consists of solutions to all problem sets and applications posed in *Angular Momentum*.

This companion book to *Angular Momentum* is intended to aid the reader to overcome difficulties encountered in learning angular momentum theory by supplying detailed illustrations of the solution methods with an emphasis on those aspects that are often not apparent. Step-by-step explanations are provided to save the reader from the large amount of time often needed to fill in the missing gaps that are usually found between the steps of illustrations in textbooks.

Our experience has been that the sharing of proposed solutions to problems often accelerates learning. Although it is possible to abuse this resource and lull oneself into a false sense of understanding, the topic of angular momentum theory may be best learned by studying solution techniques themselves. Indeed, this approach parallels closely our own experiences of how we learn new things in the laboratory.

As is normally the case in any complex undertaking, mistakes and misprints are likely to occur. Those found in the second printing of *Angular Momentum* are presented in the last chapter of this book. We have tested this material on our classes, but ultimately we accept full responsibility for whatever errors remain. We ask the reader to inform us of needed corrections so that future printings can better serve future readers.

This book project began as two independent efforts, one at the University of Illinois at Chicago, another at Stanford University, to prepare solutions for students taking courses based on *Angular Momentum* as a text. Later we combined forces to put together this book. We wish to express our sincere thanks to all those students and teaching assistants who have helped us understand better how to present this material.

Valeria Kleiman
Hongkun Park
Robert J. Gordon
Richard N. Zare

Chapter 1

ANGULAR MOMENTUM OPERATORS AND WAVE FUNCTIONS

PROBLEM SET 1

1.   DERIVE EQUATION Z-1.20[1]

We recall the definition of raising and lowering operators from equation Z-1.13,

$$j_+ = j_x + ij_y \quad \text{and} \quad j_- = j_x - ij_y. \tag{1.1}$$

It follows that

$$
\begin{aligned}
j_\mp \, j_\pm \;&=\; (j_x \mp ij_y)(j_x \pm ij_y) \\[2ex]
&=\; j_x^2 \pm ij_xj_y \mp ij_yj_x + j_y^2 \\[2ex]
&=\; j_x^2 + j_y^2 \pm i(j_xj_y - j_yj_x) \\[2ex]
&=\; j_x^2 + j_y^2 \pm i[j_x, j_y] \\[2ex]
&=\; j_x^2 + j_y^2 \mp j_z \\[2ex]
&=\; j_x^2 + j_y^2 + j_z^2 - j_z^2 \mp j_z \\[2ex]
&=\; \mathbf{j}^2 - j_z(j_z \pm 1).
\end{aligned}
\tag{1.2}
$$

2.   PROVE EQUATION Z-1.56.

From equation Z-1.43 we know that

$$Y_{\ell m}(\theta, \phi) = \Theta_{\ell m}(\theta)\, \Phi_m(\phi) \tag{1.3}$$

where

$$\Phi_m(\phi) = \frac{1}{\sqrt{2\pi}}\, e^{im\phi}. \tag{1.4}$$

Substituting equation 1.4 into 1.3 and taking its complex conjugate, we obtain

$$Y_{\ell m}^*(\theta, \phi) \;=\; [\Theta_{\ell m}(\theta)\, \Phi_m(\phi)]^*$$

---

[1]Equation numbers preceded by the letter *Z* refer to equations in *Angular Momentum: Understanding Spatial Aspects in Chemistry and Physics*, by R. N. Zare (Wiley, 1988). Here, equation Z-1.20 refers to equation 1.20 in the original text. Elsewhere an equation number such as Z-3.8.22 refers to equation 22 in Application 8, Chapter 3 of the original text.

$$= \Theta_{\ell m}(\theta) \frac{1}{\sqrt{2\pi}} e^{-im\phi}$$

$$= \Theta_{\ell m}(\theta) \, \Phi_{-m}(\phi), \tag{1.5}$$

where we have taken into account that $\Theta_{\ell m}(\theta)$ is real. From equation Z- 1.53,

$$\Theta_{\ell-|m|} = (-1)^m \Theta_{\ell|m|}, \tag{1.6}$$

it follows that for both $m > 0$ and $m < 0$,

$$\Theta_{\ell m} = (-1)^m \Theta_{\ell-m}. \tag{1.7}$$

Substituting equation 1.7 into 1.5, we get

$$Y_{\ell m}^*(\theta, \phi) = \Theta_{\ell m}(\theta) \, \Phi_{-m}(\phi)$$

$$= (-1)^m \Theta_{\ell-m}(\theta) \, \Phi_{-m}(\phi)$$

$$= (-1)^m \, Y_{\ell-m}(\theta, \phi). \tag{1.8}$$

3.   In the $|jm\rangle$ representation in which $\mathbf{j}^2$ and j_z are diagonal, the matrix elements of an arbitrary operator $\mathcal{O}$ diagonal in $\mathbf{j}^2$ are given by $\mathcal{O}_{mm'} = \langle jm | \mathcal{O} | jm' \rangle$. Each such operator may be represented by a $(2j + 1) \times (2j + 1)$ matrix where the rows are labeled by m and the columns by m'. Such matrices are called *representations*.

3A.   For the $j = 2$ case, write down explicitly the 5×5 matrices for the operators

$$j_x, \quad j_y, \quad j_z, \quad j_+, \quad j_-, \quad \mathbf{j}^2.$$

From equations Z-1.29 to Z-1.33, we get for $j = 2$

$$\langle 2m | \mathbf{j}^2 | 2m' \rangle = 6 \, \delta_{mm'}, \tag{1.9}$$

$$\langle 2m | j_z | 2m' \rangle = m' \, \delta_{mm'}, \tag{1.10}$$

$$\langle 2m | \mathrm{J}_\pm | 2m' \rangle = [6 - m'(m' \pm 1)]^{\frac{1}{2}} \, \delta_{mm'\pm 1}, \tag{1.11}$$

$$\langle 2m | j_x | 2m' \rangle = \frac{1}{2} [6 - m'(m' \pm 1)]^{\frac{1}{2}} \, \delta_{mm'\pm 1}, \tag{1.12}$$

$$\langle 2m | j_y | 2m' \rangle = \mp \frac{i}{2} [6 - m'(m' \pm 1)]^{\frac{1}{2}} \, \delta_{mm'\pm 1}. \tag{1.13}$$

Evaluating these matrix elements, we obtain

$$j_x = \begin{bmatrix} 0 & 1 & 0 & 0 & 0 \\ 1 & 0 & \sqrt{\frac{3}{2}} & 0 & 0 \\ 0 & \sqrt{\frac{3}{2}} & 0 & \sqrt{\frac{3}{2}} & 0 \\ 0 & 0 & \sqrt{\frac{3}{2}} & 0 & 1 \\ 0 & 0 & 0 & 1 & 0 \end{bmatrix}, \tag{1.14}$$

$$j_y = \begin{bmatrix} 0 & -i & 0 & 0 & 0 \\ i & 0 & -\sqrt{\frac{3}{2}}i & 0 & 0 \\ 0 & \sqrt{\frac{3}{2}}i & 0 & -\sqrt{\frac{3}{2}}i & 0 \\ 0 & 0 & \sqrt{\frac{3}{2}}i & 0 & -i \\ 0 & 0 & 0 & i & 0 \end{bmatrix}, \tag{1.15}$$

$$j_z = \begin{bmatrix} 2 & 0 & 0 & 0 & 0 \\ 0 & 1 & 0 & 0 & 0 \\ 0 & 0 & 0 & 0 & 0 \\ 0 & 0 & 0 & -1 & 0 \\ 0 & 0 & 0 & 0 & -2 \end{bmatrix}, \tag{1.16}$$

$$j_+ = \begin{bmatrix} 0 & 2 & 0 & 0 & 0 \\ 0 & 0 & \sqrt{6} & 0 & 0 \\ 0 & 0 & 0 & \sqrt{6} & 0 \\ 0 & 0 & 0 & 0 & 2 \\ 0 & 0 & 0 & 0 & 0 \end{bmatrix}, \tag{1.17}$$

$$j_- = \begin{bmatrix} 0 & 0 & 0 & 0 & 0 \\ 2 & 0 & 0 & 0 & 0 \\ 0 & \sqrt{6} & 0 & 0 & 0 \\ 0 & 0 & \sqrt{6} & 0 & 0 \\ 0 & 0 & 0 & 2 & 0 \end{bmatrix}, \tag{1.18}$$

and

$$\mathbf{j}^2 = \begin{bmatrix} 6 & 0 & 0 & 0 & 0 \\ 0 & 6 & 0 & 0 & 0 \\ 0 & 0 & 6 & 0 & 0 \\ 0 & 0 & 0 & 6 & 0 \\ 0 & 0 & 0 & 0 & 6 \end{bmatrix}. \tag{1.19}$$

3B.   By carrying out the indicated matrix operations, show that

$$j_x j_y - j_y j_x = i j_z.$$

Using the matrix representation of the angular momentum operators in the previous problem, we get

$$
j_x j_y - j_y j_x = \begin{bmatrix} 0 & 1 & 0 & 0 & 0 \\ 1 & 0 & \sqrt{\frac{3}{2}} & 0 & 0 \\ 0 & \sqrt{\frac{3}{2}} & 0 & \sqrt{\frac{3}{2}} & 0 \\ 0 & 0 & \sqrt{\frac{3}{2}} & 0 & 1 \\ 0 & 0 & 0 & 1 & 0 \end{bmatrix} \times \begin{bmatrix} 0 & -i & 0 & 0 & 0 \\ i & 0 & -\sqrt{\frac{3}{2}}i & 0 & 0 \\ 0 & \sqrt{\frac{3}{2}}i & 0 & -\sqrt{\frac{3}{2}}i & 0 \\ 0 & 0 & \sqrt{\frac{3}{2}}i & 0 & -i \\ 0 & 0 & 0 & i & 0 \end{bmatrix}
$$

$$
- \begin{bmatrix} 0 & -i & 0 & 0 & 0 \\ i & 0 & -\sqrt{\frac{3}{2}}i & 0 & 0 \\ 0 & i\sqrt{\frac{3}{2}}i & 0 & -\sqrt{\frac{3}{2}}i & 0 \\ 0 & 0 & \sqrt{\frac{3}{2}}i & 0 & -i \\ 0 & 0 & 0 & i & 0 \end{bmatrix} \times \begin{bmatrix} 0 & 1 & 0 & 0 & 0 \\ 1 & 0 & \sqrt{\frac{3}{2}} & 0 & 0 \\ 0 & \sqrt{\frac{3}{2}} & 0 & \sqrt{\frac{3}{2}} & 0 \\ 0 & 0 & \sqrt{\frac{3}{2}} & 0 & 1 \\ 0 & 0 & 0 & 1 & 0 \end{bmatrix}
$$

$$
= \begin{bmatrix} i & 0 & -\sqrt{\frac{3}{2}}i & 0 & 0 \\ 0 & \frac{1}{2}i & 0 & -\frac{3}{2}i & 0 \\ \sqrt{\frac{3}{2}}i & 0 & 0 & 0 & -\sqrt{\frac{3}{2}}i \\ 0 & \frac{3}{2}i & 0 & -\frac{1}{2}i & 0 \\ 0 & 0 & \sqrt{\frac{3}{2}}i & 0 & -i \end{bmatrix} - \begin{bmatrix} -i & 0 & -\sqrt{\frac{3}{2}}i & 0 & 0 \\ 0 & -\frac{1}{2}i & 0 & -\frac{3}{2}i & 0 \\ \sqrt{\frac{3}{2}}i & 0 & 0 & 0 & -\sqrt{\frac{3}{2}}i \\ 0 & \frac{3}{2}i & 0 & \frac{1}{2}i & 0 \\ 0 & 0 & \sqrt{\frac{3}{2}}i & 0 & i \end{bmatrix}
$$

$$
= i \begin{bmatrix} 2 & 0 & 0 & 0 & 0 \\ 0 & 1 & 0 & 0 & 0 \\ 0 & 0 & 0 & 0 & 0 \\ 0 & 0 & 0 & -1 & 0 \\ 0 & 0 & 0 & 0 & -2 \end{bmatrix}
$$

$$
= i\, j_z. \tag{1.20}
$$

3C. FIND THE SPECIFIC MATRIX ELEMENT $\langle j = 2, m = 0 | j_x j_y j_x | j = 2, m = 1 \rangle$.

From the matrix representations of j_x and $j_x j_y$, we obtain

$$
\langle j = 2, m = 0 | j_x j_y j_x | j = 2, m = 1 \rangle = \langle 20 | j_x j_y j_x | 21 \rangle
$$

$$
= \sum_{m=-2}^{2} \langle 20 | j_x j_y | 2m \rangle \langle 2m | j_x | 21 \rangle
$$

$$= \left[\begin{array}{ccccc} \sqrt{\frac{3}{2}}i & 0 & 0 & 0 & -\sqrt{\frac{3}{2}}i \end{array} \right] \left[\begin{array}{c} 1 \\ 0 \\ \sqrt{\frac{3}{2}} \\ 0 \\ 0 \end{array} \right]$$

$$= \sqrt{\frac{3}{2}}\,i, \tag{1.21}$$

where we have used the identity matrix $\sum_m |2m\rangle\langle 2m| = \underset{\sim}{I}$.

4.   We explore an alternative procedure for identifying $\langle \mathbf{j}^2 \rangle = \langle jm|\mathbf{j}^2|jm \rangle$ with $j(j+1)$. This is accomplished by equating the expectation value of an operator $\mathcal{O}$ with its spatial average, defined by

$$(\mathcal{O})_{sp} = \frac{\sum_m \langle jm|\mathcal{O}|jm \rangle}{\sum_m \langle jm|jm \rangle}. \tag{1.22}$$

Because the choice of coordinates is arbitrary and space is isotropic,

$$\langle \mathbf{j}^2 \rangle = \langle j_x \rangle^2 + \langle j_y \rangle^2 + \langle j_z \rangle^2$$

$$= 3(j_z^2)_{sp}. \tag{1.23}$$

Perform the spatial average and show that $\langle \mathbf{j}^2 \rangle = j(j+1)$.

With the help of equation 1.22 we evaluate $(j_z^2)_{sp}$,

$$(j_z^2)_{sp} = \frac{\sum_m \langle jm|j_z^2|jm \rangle}{\sum_m \langle jm|jm \rangle}$$

$$= \frac{\sum_m m^2}{\sum_m 1}$$

$$= \frac{1}{3}\frac{j(j+1)(2j+1)}{(2j+1)}$$

$$= \frac{1}{3}j(j+1), \tag{1.24}$$

where we have used the normalization $\langle jm|jm \rangle = 1$ and $\langle jm \left| j_z^2 \right| jm \rangle = m^2$. Substituting equation 1.24 into equation 1.23 gives

$$\langle \mathbf{j}^2 \rangle = j(j+1). \tag{1.25}$$

APPLICATION 1
SCATTERING THEORY

A.

SHOW THAT

$$L_z = \mu r^2 \dot{\theta}$$

AND

$$E = \frac{1}{2}\mu(\dot{r}^2 + r^2\dot{\theta}^2) + V(r).$$

Hint: INTRODUCE POLAR COORDINATES

$$x = r\cos\theta$$

$$y = r\sin\theta$$

TO DESCRIBE THE MOTION IN THE SCATTERING PLANE AND EVALUATE

$$L = \mu(x\dot{y} - y\dot{x})$$

$$E = \frac{1}{2}\mu(\dot{x}^2 + \dot{y}^2) + V(r).$$

Because $V = V(r)$, $\mathbf{L}$ is a constant of the motion. If we choose the z axis of our coordinate system along the direction of $\mathbf{L}$, then

$$\mathbf{L} = L_z\hat{z}$$

and

$$L_z = \mu(x\dot{y} - y\dot{x}).$$

Using polar coordinates, we get

$$
\begin{aligned}
L_z &= \mu(r\cos\theta\frac{\partial(r\sin\theta)}{\partial t} - \frac{\partial(r\cos\theta)}{\partial t}r\sin\theta) \\[2mm]
&= \mu r[\cos\theta(\dot{r}\sin\theta + r\dot{\theta}\cos\theta) - \sin\theta(\dot{r}\cos\theta - r\dot{\theta}\sin\theta)] \\[2mm]
&= \mu r(r\dot{\theta}\cos^2\theta + r\dot{\theta}\sin^2\theta) \\[2mm]
&= \mu r^2\dot{\theta}.
\end{aligned}
$$

$$(1.26)$$

Because the motion of the mass point is confined to the xy plane, we may express the energy as

$$E = \frac{1}{2}\mu(\dot{x}^2 + \dot{y}^2) + V(r)$$

$$= \frac{1}{2}\mu\left[\left(\frac{\partial(r\cos\theta)}{\partial t}\right)^2 + \left(\frac{\partial(r\sin\theta)}{\partial t}\right)^2\right] + V(r)$$

$$= \frac{1}{2}\mu[(\dot{r}\cos\theta - r\dot{\theta}\sin\theta)^2 + (\dot{r}\sin\theta + r\dot{\theta}\cos\theta)^2] + V(r)$$

$$= \frac{1}{2}\mu(\dot{r}^2 + r^2\dot{\theta}^2) + V(r). \tag{1.27}$$

B.

DERIVE

$$b^2 = r_c^2\left(1 - \frac{V(r_c)}{E}\right)$$

FROM

$$E = \frac{(\mu v b)^2}{2\mu r_c^2} + V(r_c),$$

WHERE r_c IS THE DISTANCE OF CLOSEST APPROACH.

In the asymptotic limit where $\dot{r} = v$,

$$E = \frac{1}{2}\mu v^2,$$

and

$$L = \mu v b.$$

At closest approach, $r = r_c$ and $\dot{r} = 0$. Thus we obtain from equations 1.26 and 1.27

$$E = \frac{(\mu v b)^2}{2\mu r_c^2} + V(r_c)$$

$$= \frac{1}{2}\mu v^2\, \frac{b^2}{r_c^2} + V(r_c)$$

$$= E\, \frac{b^2}{r_c^2} + V(r_c), \tag{1.28}$$

which can be rearranged to yield

$$b^2 = r_c^2 \left(1 - \frac{V(r_c)}{E}\right).$$

(1.29)

C.

CONSIDER A SPHERE OF RADIUS a, SUCH THAT

$$V(r) = \infty \quad \text{FOR} \quad r < a,$$

AND

$$V(r) = 0 \quad \text{FOR} \quad r > a.$$

SHOW THAT THE CLASSICAL DIFFERENTIAL CROSS SECTION IS INDEPENDENT OF THE DEFLECTION ANGLE AND THAT THE CLASSICAL TOTAL CROSS SECTION IS THE AREA OF A CIRCLE OF RADIUS a. *Hint*: BEGIN BY PROVING THAT $\chi = 2 \arccos \frac{b}{a}$ FOR $b \leq a$, $\chi = 0$ FOR $b > a$.

Because the potential is null for $r > a$, there is no interaction in that region and, therefore, $\chi = 0$ for $b > a$. For $r \leq a$, the potential goes to infinity, and the particle cannot penetrate into the sphere. It follows that for $b < a$ the trajectories consist of two straight lines symmetrical about the radial vector at the point of impact (see Figure Z-1.1). The point of closest approach is the radius of the sphere itself, and the scattering angle is given by

$$\chi = \pi - 2b \int_{r_c=a}^{\infty} \frac{\mathrm{d}r}{r^2 \left[1 - \frac{b^2}{r^2} - \frac{V(r)}{E}\right]^{\frac{1}{2}}}.$$

(1.30)

Case (i): If $b \leq a$, then $r_c = a$, and the deflection angle is given by

$$\chi = \pi - 2b \lim_{\varepsilon \to 0} \int_{a+\varepsilon}^{\infty} \frac{\mathrm{d}r}{r^2 \left[1 - \frac{b^2}{r^2} - \frac{V(a+\varepsilon)}{E}\right]^{\frac{1}{2}}}.$$

(1.31)

Because $V(a + \varepsilon) = 0$, we get

$$\chi = \pi - 2b \lim_{\varepsilon \to 0} \int_{a+\varepsilon}^{\infty} \frac{\mathrm{d}r}{r^2 \left[1 - \frac{b^2}{r^2}\right]^{\frac{1}{2}}}.$$

(1.32)

Setting $r = \frac{b}{x}$, we obtain

$$\chi = \pi - 2b \lim_{\varepsilon \to 0} \int_{\frac{b}{a+\varepsilon}}^{\infty} \left(-\frac{1}{b}\right) \frac{\mathrm{d}x}{\sqrt{1 - x^2}}$$

$$= \pi - 2b \lim_{\varepsilon \to 0} \frac{1}{b} \arccos x \Big|_{\frac{b}{a+\varepsilon}}^{0},$$

(1.33)

and

$$\chi_{b \leq a} = 2 \arccos \frac{b}{a}.$$

(1.34)

Case (ii): If $b > a$, then $r_c = b$, and the deflection angle is given by

$$\chi = \pi - 2b \int_b^\infty \frac{\mathrm{d}r}{r^2 \left[1 - \frac{b^2}{r^2} - \frac{V(b)}{E}\right]^{\frac{1}{2}}} \ . \tag{1.35}$$

Because $V(b) = 0$, we obtain

$$\chi = \pi - 2b \int_b^\infty \frac{\mathrm{d}r}{r^2 \sqrt{1 - \frac{b^2}{r^2}}} \ . \tag{1.36}$$

Setting $r = \frac{b}{x}$, we get

$$\chi = \pi - 2b \int_1^0 \left(-\frac{1}{b}\right) \frac{\mathrm{d}x}{\sqrt{1 - x^2}} \tag{1.37}$$

and

$$\chi_{b>a} = 0. \tag{1.38}$$

We can now use equations 1.34 and 1.38 to calculate the differential cross section,

$$I(\chi) = \frac{b}{\sin\chi \left|\frac{\mathrm{d}\chi}{\mathrm{d}b}\right|} \ . \tag{1.39}$$

For $b > a$, because the particle doesn't "feel" any field, the deflection angle is zero. For $b \leq a$, the deflection angle is given by equation 1.34. Differentiating equation 1.34, we obtain

$$\left|\frac{\partial\chi}{\partial b}\right| = \left|\frac{-2}{a\sin\frac{\chi}{2}}\right| . \tag{1.40}$$

Substituting equation 1.40 into equation 1.39 and using the trigonometric identity $\sin\chi = 2\sin\frac{\chi}{2}\cos\frac{\chi}{2}$, we obtain

$$\begin{aligned}
I(\chi) &= \frac{b}{2\sin\frac{\chi}{2}\cos\frac{\chi}{2}\left|\frac{\mathrm{d}\chi}{\mathrm{d}b}\right|} \\[2mm]
&= \frac{ba}{4\cos\frac{\chi}{2}} \\[2mm]
&= \frac{a^2}{4} \ . \tag{1.41}
\end{aligned}$$

Equation 1.41 shows that $I(\chi)$ is independent of the angle of deflection χ and depends only on the radius of the sphere. The total cross section σ is given by

$$\begin{aligned}
\sigma &= 2\pi \int_0^\pi I(\chi) \sin\chi \, d\chi \\[2em]
&= 2\pi \frac{a^2}{4} \int_0^\pi \sin\chi \, d\chi \\[2em]
&= \pi a^2.
\end{aligned}$$

(1.42)

The total cross section is just the cross sectional area of the hard sphere.

D.

LET US EXAMINE THE SMALL-ANGLE SCATTERING FROM THE POTENTIAL FUNCTION $V(r) = C_s r^{-s}$ WITH $s > 0$.

We know that

$$\chi = -\frac{b}{E} \int_b^\infty \left(\frac{\partial V}{\partial r}\right) \left(r^2 - b^2\right)^{-\frac{1}{2}} \, dr,$$

(1.43)

and we calculate that

$$\frac{\partial V}{\partial r} = -s \, C_s \, r^{-(s+1)}.$$

(1.44)

Substituting equation 1.44 into equation 1.43, we obtain

$$\begin{aligned}
\chi &= \frac{bsC_s}{E} \int_b^\infty r^{-(s+1)} \left(r^2 - b^2\right)^{-\frac{1}{2}} \, dr \\[2em]
&= \frac{bsC_s}{E} \int_b^\infty r^{-(s+1)} \left(1 - \frac{b^2}{r^2}\right)^{-\frac{1}{2}} \frac{dr}{r}.
\end{aligned}$$

(1.45)

To evaluate the integral, we make the substitutions

$$u = \frac{b^2}{r^2} \qquad \text{and} \qquad du = -\frac{2b^2}{r^3} \, dr,$$

(1.46)

and the integral becomes

$$\begin{aligned}
\chi &= \frac{bsC_s}{2E} \int_0^1 b^{-(s+1)} u^{\frac{s+1}{2}} \left(1 - u\right)^{-\frac{1}{2}} \frac{du}{u} \\[2em]
&= \frac{sC_s}{2Eb^s} \int_0^1 u^{\frac{s-1}{2}} \left(1 - u\right)^{-\frac{1}{2}} du
\end{aligned}$$

$$= \frac{sC_s}{2Eb^s} \frac{\Gamma(\frac{s+1}{2})\Gamma(\frac{1}{2})}{\Gamma(\frac{s}{2}+1)}$$

$$= \frac{sC_s\sqrt{\pi}}{2Eb^s} \frac{\Gamma(\frac{s+1}{2})}{\Gamma(\frac{s}{2}+1)}, \tag{1.47}$$

where we have used $\int_0^1 u^{m-1}(1-u)^{n-1}\,du = \frac{\Gamma(m)\Gamma(n)}{\Gamma(m+n)}$ and $\Gamma(\frac{1}{2}) = \sqrt{\pi}$.
We can now evaluate $I(\chi)$ from equation 1.39. Because

$$\left|\frac{\partial\chi}{\partial b}\right| = \left|\frac{sC_s\sqrt{\pi}}{2E} \times \frac{\Gamma(\frac{s+1}{2})}{\Gamma(\frac{s}{2}+1)} \times \frac{s}{b^{s+1}}\right| \tag{1.48}$$

and $\sin\chi \approx \chi$ for small χ, it follows that

$$I(\chi) = (b \times b^{s+1} \times b^s)\left(\frac{s^2C_s\sqrt{\pi}\Gamma(\frac{s+1}{2})}{2E\,\Gamma(\frac{s}{2}+1)}\right)^{-1}\left(\frac{sC_s\sqrt{\pi}\Gamma(\frac{s+1}{2})}{2E\,\Gamma(\frac{s}{2}+1)}\right)^{-1}. \tag{1.49}$$

From equation 1.47 we get

$$b = \left(\frac{sC_s\sqrt{\pi}\Gamma(\frac{s+1}{2})}{2E\,\Gamma(\frac{s}{2}+1)}\right)^{\frac{1}{s}}\frac{1}{\chi^{\frac{1}{s}}}. \tag{1.50}$$

Substituting equation 1.50 into equation 1.49, we get

$$I(\chi) = \left(\frac{sC_s\sqrt{\pi}\Gamma(\frac{s+1}{2})}{2E\,\Gamma(\frac{s}{2}+1)}\right)^{\frac{2}{s}}\frac{1}{s}\frac{1}{\chi^{2+\frac{2}{s}}}$$

$$= F(s,E)\,\chi^{-(2+\frac{2}{s})}. \tag{1.51}$$

It follows that a log-log plot of $I(\chi)$ vs χ for a specific energy can be used to determine s.

E. ·

IN QUANTUM MECHANICS ONLY INTEGRAL VALUES OF ℓ ARE ALLOWED, YET WE VIEW b AS BEING CONTINUOUS. HOW WILL THIS AFFECT OUR RESULTS WHEN TREATING CHEMICAL SYSTEMS?

Classically, the angular momentum and the impact parameter are related to each other by $b \approx (\ell + \frac{1}{2})\frac{\lambda}{2\pi}$. We may picture the incident beam as consisting of a set of concentric cylinders, each one corresponding to a specific value of ℓ. The separation between these cylinders is $\frac{\lambda}{2\pi}$. The impact parameter will appear to be continuous if this separation is much smaller than the target, whereas for a quantum system, the separation is comparable with the size of the target.

To illustrate this, we calculate $\frac{\lambda}{2\pi}$ for two cases. For a Maxwell-Boltzmann distribution, the average momentum carried by a particle of mass m is

$$p = \sqrt{\frac{8mkT}{\pi}}, \tag{1.52}$$

giving

$$\frac{\lambda}{2\pi} = h(32\pi mkT)^{-\frac{1}{2}}. \tag{1.53}$$

For C_2H_5OH, $m = 7.66 \times 10^{-26}$ kg. At room temperature, equation 1.53 gives $\frac{\lambda}{2\pi} = 0.037$Å. Because the Bohr radius is 0.53Å, we see that for a typical molecule the impact parameter is effectively continuous. Doing the same calculation for H, $\frac{\lambda}{2\pi} = 0.25$Å. In this case, $\frac{\lambda}{2\pi}$ is comparable with the Bohr radius. We expect that the quantization of ℓ will be important for collisions involving light atoms, and it will be even more pronounced for electrons!

F.

DERIVE

$$f(\chi) = -\frac{\lambda}{2\pi} \sum_{\ell=0}^{\infty} \sqrt{\frac{\ell + \frac{1}{2}}{2\pi \; \sin \chi}} \left(e^{i\phi^+} - e^{i\phi^-} \right).$$

Hint: REREAD NOTE 8 OF CHAPTER 1.

Using the partial-wave expansion, the scattering amplitude is given by

$$f(\chi) = \frac{1}{2i} \frac{\lambda}{2\pi} \sum_{\ell=0}^{\infty} (2\ell + 1) \; e^{2i\eta_\ell} \; P_\ell(\cos \chi). \tag{1.54}$$

To derive the classical limit of the amplitude, we consider a slowly varying potential and a contribution from large values of ℓ to the scattering angle. For ℓ large enough that $\chi \gg \ell^{-1}$,

$$P_\ell(\cos \chi) = \sqrt{\frac{2}{(\ell + \frac{1}{2}) \; \pi}} \frac{\sin \left[(\ell + \frac{1}{2})\chi + \frac{\pi}{4} \right]}{(\sin \chi)^{\frac{1}{2}}}. \tag{1.55}$$

Substituting equation 1.55 into equation 1.54, we obtain

$$
\begin{aligned}
f(\chi) &= \frac{1}{2i} \frac{\lambda}{2\pi} \sum_{\ell=0}^{\infty} (2\ell + 1) \; e^{2i\eta_\ell} \sqrt{\frac{2}{(\ell + \frac{1}{2}) \; \pi}} \frac{\sin \left[(\ell + \frac{1}{2})\chi + \frac{\pi}{4} \right]}{(\sin \chi)^{\frac{1}{2}}} \\[2ex]
&= -i \frac{\lambda}{2\pi} \sum_{\ell=0}^{\infty} \sqrt{\frac{2\ell + 1}{\pi \sin \chi}} e^{2i\eta_\ell} \left[-\frac{i}{2} \left(e^{i[(\ell+\frac{1}{2})\chi + \frac{\pi}{4}]} - e^{-i[(\ell+\frac{1}{2})\chi + \frac{\pi}{4}]} \right) \right] \\[2ex]
&= -\frac{\lambda}{2\pi} \sum_{\ell=0}^{\infty} \sqrt{\frac{\ell + \frac{1}{2}}{2\pi \; \sin \chi}} \left(e^{i\phi^+} - e^{i\phi^-} \right),
\end{aligned} \tag{1.56}
$$

where $\phi^{\pm} = 2\eta_\ell \pm (\ell + \frac{1}{2})\chi \pm \frac{\pi}{4}$.

G.

THE CROSS SECTION THAT EACH PARTIAL WAVE ℓ CONTRIBUTES TO THE TOTAL CROSS SECTION IS GIVEN BY

$$\sigma_\ell = \frac{4\pi}{k^2}(2\ell + 1)\sin^2 \eta_\ell. \tag{1.57}$$

GENERALLY THE PHASE SHIFTS η_ℓ ARE SLOWLY VARYING FUNCTIONS OF THE ENERGY. A RESONANCE OCCURS FOR SOME PARTIAL WAVE ℓ WHEN η_ℓ CHANGES RAPIDLY OVER A SMALL ENERGY RANGE, IN WHICH THE PHASE SHIFT MAY BE DIVIDED INTO A BACKGROUND PART AND A RESONANCE PART,

$$\eta_\ell(k) = \eta_{bg} + \eta_{res}, \tag{1.58}$$

WHERE

$$\eta_{res} = \arctan\left(\frac{\Gamma/2}{E_0 - E}\right). \tag{1.59}$$

SHOW THAT THE PARTIAL CROSS SECTION IS THEN GIVEN BY

$$\sigma_\ell = \frac{4\pi}{k^2}(2\ell + 1)\sin^2\left[\eta_{bg} + \arctan\left(\frac{\Gamma/2}{E_0 - E}\right)\right]$$

$$= \frac{4\pi}{k^2}(2\ell + 1)\frac{(\varepsilon + q)^2}{1 + \varepsilon^2}\sin^2 \eta_{bg},$$

WHERE

$$\varepsilon = \frac{E_0 - E}{\Gamma/2}$$

IS THE ENERGY DIFFERENCE MEASURED IN HALF-WIDTHS, AND

$$q = \cot \eta_{bg}$$

IS CALLED THE *line profile* INDEX.

Defining $\beta = \frac{\Gamma/2}{E_0 - E}$, it follows that

$$\sigma_\ell = \frac{4\pi}{k^2}(2\ell + 1)\sin^2\left[\eta_{bg} + \arctan\beta\right]$$

$$= \frac{4\pi}{k^2}(2\ell + 1)\left[\sin\eta_{bg}\ \cos(\arctan\beta) + \cos\eta_{bg}\ \sin(\arctan\beta)\right]^2$$

$$= \frac{4\pi}{k^2}(2\ell + 1)\left[\sin^2\eta_{bg}\ \cos^2(\arctan\beta) + \cos^2\eta_{bg}\ \sin^2(\arctan\beta)\right.$$

$$\left. + 2\sin(\arctan\beta)\ \cos(\arctan\beta)\ \sin\eta_{bg}\ \cos\eta_{bg}\right]$$

$$= \frac{4\pi}{k^2}(2\ell + 1)\frac{1}{1 + \beta^2}\left(\sin^2\eta_{bg} + \beta^2\cos^2\eta_{bg} + 2\beta\ \sin\eta_{bg}\ \cos\eta_{bg}\right), \tag{1.60}$$

where we have used the identity

$$\cos^2(\arctan\beta) = 1/(1+\beta^2). \tag{1.61}$$

We can further show that

$$\sin^2\eta_{bg} + \beta^2\,\cos^2\eta_{bg} + 2\beta\,\sin\eta_{bg}\,\cos\eta_{bg} = \left(\frac{1}{\beta}+cot\eta_{bg}\right)^2\beta^2\,\sin^2\eta_{bg}. \tag{1.62}$$

Considering the definitions of ε as the energy difference at half-width and q as the line profile index, we obtain from equation 1.60

$$\sigma_\ell = \frac{4\pi}{k^2}(2\ell+1)\,(\varepsilon+q)^2\sin\eta_{bg}{}^2\frac{\beta^2}{1+\beta^2}$$

$$= \frac{4\pi}{k^2}(2\ell+1)\frac{(\varepsilon+q)^2}{1+\varepsilon^2}\sin^2\eta_{bg}. \tag{1.63}$$

Chapter 2

COUPLING OF TWO ANGULAR MOMENTUM VECTORS

APPLICATION 2
THE WIGNER-WITMER RULES

A.

SHOW THAT THERE ARE

$$(2S_< + 1)(2L_< + 1)$$

Σ TERMS (CALLED Σ *states*) OF DIFFERENT MULTIPLICITY.

Σ terms occur when $\Lambda = |M_1 + M_2| = 0$. For each combination of L and S, this occurs only once. For each value of S, there are a total of $2L_< + 1$ values of L, namely,

$$L = L_> - L_<, \ L_> - L_< + 1, \ldots, L_> + L_<.$$

The total number of values of S is $2S_< + 1$, corresponding to

$$S = S_> - S_<, \ S_> - S_< + 1, \ldots, \ S_> + S_<.$$

The total number of Σ terms is therefore

$$N_\Sigma = (2S_< + 1)(2L_< + 1). \tag{2.1}$$

B.

SHOW THAT THERE ARE A TOTAL OF

$$(2S_< + 1)(2L_< + 1)(L_> + 1)$$

ELECTRONIC TERMS, WHERE Λ RANGES FROM 0 TO $L_1 + L_2$ AND S RANGES FROM $|S_1 - S_2|$ TO $S_1 + S_2$, THAT IS, FROM $S_> - S_<$ TO $S_> + S_<$.

Each pair of values of M_{L_1} and M_{L_2} produces a distinct Λ state, where

$$\Lambda = |M_{L_1} + M_{L_2}|. \tag{2.2}$$

Without loss of generality, we may choose $L_1 < L_2$ such that $M_{L_1} = M_{L_<}$ and $M_{L_2} = M_{L_>}$. Because $M_{L_<}$ ranges from $-L_<$ to $+L_<$ and $M_{L_>}$ from $-L_>$ to $+L_>$, there are

$$(2L_< + 1)(2L_> + 1) \tag{2.3}$$

Λ states. The next step is to take into account the degeneracy of states with $\Lambda > 0$. First we subtract the number of Σ states, then we divide by the degeneracy factor of 2, and finally we add back the Σ states. For each spin state, the number of non-Σ states is

$$(2L_< + 1)(2L_> + 1) - (2L_< + 1) = 2L_> (2L_< + 1). \tag{2.4}$$

The number of total L terms is then

$$\frac{2L_> (2L_< + 1)}{2} + (2L_< + 1) = (2L_< + 1)(L_> + 1). \tag{2.5}$$

Multiplying by the number of spin states, we get

$$N = (2S_< + 1)(2L_< + 1)(L_> + 1) \tag{2.6}$$

for the total number of electronic terms.

C.

UNDER INVERSION $x \to -x$, $y \to -y$, $z \to -z$, SHOW THAT $\phi \to \pi + \phi$ AND $\theta \to \pi - \theta$.

In spherical coordinates,

$$x = r \sin\theta \cos\phi,$$

$$y = r \sin\theta \sin\phi,$$

$$z = r \cos\theta. \tag{2.7}$$

Under inversion we get

$$x' = -x = -r \sin\theta \cos\phi = r \sin\theta' \cos\phi',$$

$$y' = -y = -r \sin\theta \sin\phi = r \sin\theta' \sin\phi',$$

$$z' = -z = -r \cos\theta = r \cos\theta'. \tag{2.8}$$

From the inversion of z, we learn that $\cos\theta' = -\cos\theta$ and $\theta' = \pi - \theta$, where θ is defined in the domain $(0, \pi)$. Substituting this result into the inversion of x and y, we get

$$\sin\theta' = \sin(\pi - \theta) = \sin\theta,$$

which can be used to obtain the relation between ϕ and ϕ',

$$\left. \begin{array}{rcl} \cos\phi' & = & -\cos\phi \\ \sin\phi' & = & -\sin\phi \end{array} \right\} \phi' = \pi + \phi. \tag{2.9}$$

D.

USE THE PRECEDING RESULT TO PROVE THAT THE SPHERICAL HARMONICS UNDER INVERSION OBEY THE RELATION

$$\hat{\imath} Y_{\ell m}(\theta, \phi) = (-1)^{\ell} Y_{\ell m}(\theta, \phi), \tag{2.10}$$

FROM WHICH IT IS CONCLUDED THAT THE PARITY OF THE STATE OF A PARTICLE WITH ANGULAR MOMENTUM ℓ IS INDEPENDENT OF m. *Hint*: SHOW THAT UNDER THIS TRANSFORMATION $\theta \to \pi - \theta$, AND $\cos\theta \to -\cos\theta$.

The spherical harmonics can be written as a product of two functions,

$$Y_{\ell m}(\theta, \phi) = \Theta_{\ell m}(\theta) \, \Phi_m(\phi). \tag{2.11}$$

For $m \geq 0$, we know that

$$\Theta_{\ell m}(\theta) = (-1)^m \times constant \times P_L^M(\cos\theta). \tag{2.12}$$

We need to show how $P_L^M(\cos\theta)$ changes with $\theta \to \pi - \theta$ and how $\Phi_m(\phi) = e^{im\phi}$ changes with $\phi \to \pi + \phi$. The latter transformation yields

$$
\begin{aligned}
e^{im\phi'} &= e^{im(\pi+\phi)} \\[2ex]
&= e^{im\pi}\, e^{im\phi} \\[2ex]
&= (-1)^m\, e^{im\phi},
\end{aligned}
\tag{2.13}
$$

which gives

$$
\Phi_m(\phi') = (-1)^m\, \Phi_m(\phi).
\tag{2.14}
$$

Because $\cos\theta' = -\cos\theta$,

$$
\begin{aligned}
P_L^M(-\cos\theta) &= \frac{\sin^m(\pi-\theta)}{2^\ell \ell!} \left[\frac{\partial}{\partial(-\cos\theta)}\right]^{m+\ell} \left[\cos^2(\pi-\theta) - 1\right]^\ell \\[3ex]
&= \frac{\sin^m(\theta)}{2^\ell \ell!} \left[\frac{\partial}{\partial(\cos\theta)} \frac{\partial\cos\theta}{\partial(-\cos\theta)}\right]^{m+\ell} \left[(-\cos\theta)^2 - 1\right]^\ell \\[3ex]
&= \frac{\sin^m(\theta)}{2^\ell \ell!}(-1)^{m+\ell} \left[\frac{\partial}{\partial(\cos\theta)}\right]^{m+\ell} \left[\cos^2\theta - 1\right]^\ell.
\end{aligned}
\tag{2.15}
$$

Therefore, we obtain

$$
P_L^M(\cos\theta') = (-1)^{m+\ell} P_L^M(\cos\theta).
\tag{2.16}
$$

and

$$
\Theta_{\ell m}(\theta') = (-1)^{m+\ell}\Theta_{\ell m}(\theta).
\tag{2.17}
$$

Substituting equations 2.14 and 2.17 into equation 2.11, we get

$$
\begin{aligned}
\hat{\imath} Y_{\ell m}(\theta,\phi) &= (-1)^m (-1)^{m+\ell} Y_{\ell m}(\theta,\phi) \\[2ex]
&= (-1)^\ell Y_{\ell m}(\theta,\phi).
\end{aligned}
\tag{2.18}
$$

For $m \leq 0$, $\Theta_{\ell,|m|}$ is defined by

$$
\Theta_{\ell,-|m|} = (-1)^m \Theta_{\ell,|m|}.
\tag{2.19}
$$

Because there is no change in m under inversion, the transformation for $m \geq 0$ is also valid for $m \leq 0$.

E.

$$p = (-1)^{\Sigma_i \ell_i},$$

A rough approximation is to treat the electronic wave function as a product of single electron wave functions,

$$\Psi(1, 2, \ldots, n) = |\psi(1)\psi(2) \ldots \psi(n)|, \tag{2.20}$$

where the two vertical bars represent the Slater determinant. Each wave function may be factored into radial and angular functions,

$$\psi(n) = R_{n\ell}(r) Y_{\ell m}(\theta, \phi). \tag{2.21}$$

Applying the inversion operator to $R_{n\ell}(r)$, we obtain

$$\hat{i} R_{n\ell}(r) = R_{n\ell}(r). \tag{2.22}$$

Substituting equation 2.18 and 2.22 into equation 2.21, we obtain

$$\hat{i}\psi(n) = (-1)^{\ell}\psi(n), \tag{2.23}$$

which can be used to give

$$\hat{i}\Psi(1, 2, \ldots, n) = (-1)^{\ell_1}(-1)^{\ell_2} \ldots (-1)^{\ell_n} |\psi(1)\psi(2) \ldots \psi(n)|$$

$$= (-1)^{\Sigma_i \ell_i} \Psi(1, 2, \ldots, n). \tag{2.24}$$

Thus, the parity of the atomic wave function is given by

$$p = (-1)^{\Sigma_i \ell_i}. \tag{2.25}$$

F.

$$C_2(y)|L0\rangle = (-1)^L |L0\rangle. \tag{2.26}$$

Hint: Use the fact that $|L0\rangle$ is proportional to $P_L(\cos\theta)$.

First we see how rotation transforms θ and ϕ. (This problem was previously solved for inversion in part C). Under $C_2(y)$ rotation, we get

$$x' = -x = r \sin\theta' \cos\phi',$$

$$y' = y = r \sin\theta \sin\phi,$$

$$z' = -z = r \cos\theta'. \tag{2.27}$$

From $C_2(y)$ operating on z, we obtain $\cos\theta' = \cos\theta$ and $\theta' = \pi - \theta$ in the domain of $\theta \in (0, \pi)$. Knowing that $\sin\theta' = \sin\theta$, we can obtain the relation between ϕ and ϕ',

$$\left.\begin{array}{rcl} \cos\phi' &=& -\cos\phi \\ \sin\phi' &=& \sin\phi \end{array}\right\} \phi' = \pi - \phi.$$

Now we can evaluate how the spherical harmonics transform under $C_2(y)$ rotation (as we did in part D):

$$C_2(y)P_L^M(\cos\theta) = P_L^M(\cos(\pi - \theta))$$

$$= (-1)^{M+L}P_L^M(\cos\theta), \tag{2.28}$$

where we have used equation 2.16, and

$$e^{iM\phi'} = (-1)^M e^{-iM\phi}. \tag{2.29}$$

Substituting equations 2.28 and 2.29 into equation 2.11 with $M = 0$, we obtain

$$C_2(y)Y_{L0}(\theta, \phi) = (-1)^L Y_{L0}(\theta, \phi). \tag{2.30}$$

Because

$$|L0\rangle = R_{nL}(r)Y_{L0}(\theta, \phi), \tag{2.31}$$

we obtain

$$C_2(y)|L0\rangle = (-1)^L |L0\rangle. \tag{2.32}$$

G.

THE GROUND STATE OF THE HYDROGEN ATOM IS

$$\text{H: } 1s \ \ ^2S,$$

AND THAT OF THE OXYGEN ATOM IS

$$\text{O: } (1s)^2(2s)^2(2p)^4 \ \ ^3P.$$

FIND ALL THE ELECTRONIC STATES OF H_2, OH, AND O_2 THAT CAN BE BUILT UP BY BRINGING TOGETHER THEIR RESPECTIVE GROUND-STATE ATOMS.

We can use equation 2.1 to calculate the number of Σ states and equation 2.6 to calculate the total number of states in each case.

H_2:

$$^2S + {}^2S$$

$$\Lambda = |M_{L_1} + M_{L_2}| = 0$$

Only Σ states are possible. From part A, we know that the number of terms is

$$N = (2 \times \frac{1}{2} + 1)(2 \times 0 + 1) = 2. \tag{2.33}$$

These correspond to

$$S_1 - S_2 = 0 \quad and \quad S_1 + S_2 = 1. \tag{2.34}$$

From the rules in the text, we recognize that the first term has $+$ and g symmetry, whereas the second one has $+$ and u symmetry. The resulting terms are therefore

$$^1\Sigma_g^+ \qquad ^3\Sigma_u^+.$$

O_2:

$$^3P + {}^3P$$

$$\Lambda = 0, 1, 2$$

Σ, Π, and Δ states are possible, and the total number of terms is

$$N = (2 \times 1 + 1)(2 \times 1 + 1)(1 + 1) = 18.$$

Of these, the number of Σ states is

$$N_\Sigma = (2 \times 1 + 1)(2 \times 1 + 1) = 9.$$

The possible values of S are

$$S = 0, 1, 2.$$

For the Σ states, because $(-1)^{L_1 + L_2} = 1$, there will be one more Σ^+ state for each multiplicity. Using again the rules in the text, we identify the symmetries of these terms as follows:

$$
\begin{array}{cccccc}
^1\Sigma_g^+ & ^1\Sigma_g^+ & ^1\Sigma_u^- & ^1\Pi_g & ^1\Pi_u & ^1\Delta_g \\
^3\Sigma_u^+ & ^3\Sigma_u^+ & ^3\Sigma_g^- & ^3\Pi_g & ^3\Pi_u & ^3\Delta_u \\
^5\Sigma_g^+ & ^5\Sigma_g^+ & ^5\Sigma_u^- & ^5\Pi_g & ^5\Pi_u & ^5\Delta_g.
\end{array}
$$

OH:

$$^3P + {}^2S$$

$$\Lambda = 0, 1$$

Σ and Π states are possible, and the total number of states is given by

$$N = (2 \times 0 + 1)(2 \times \frac{1}{2} + 1)(1 + 1) = 4.$$

The number of Σ states is

$$N_\Sigma = (2 \times 0 + 1)(2 \times \frac{1}{2} + 1) = 2,$$

and the possible values of S are

$$S = \frac{1}{2}, \frac{3}{2}.$$

For the Σ states $(-1)^1 = -1$ and, therefore, they are all Σ^- states. The resulting terms are

$$
\begin{array}{cc}
^2\Sigma^- & ^2\Pi \\
^4\Sigma^- & ^4\Pi.
\end{array}
$$

APPLICATION 3
THE ROTATIONAL ENERGY LEVELS
OF A $^2\Sigma$ FREE RADICAL

A.

USE THE COUPLED REPRESENTATION $|\,NSJM\,\rangle$ TO CALCULATE EXPLICIT EXPRESSIONS IN TERMS OF THE ROTATIONAL QUANTUM NUMBER N FOR THE ENERGIES OF THE ROTATIONAL LEVELS $J = N + \frac{1}{2}$ AND $J = N - \frac{1}{2}$. HENCE SHOW THAT LEVELS WITH THE SAME VALUE OF N ARE SPLIT (DOUBLED) BY SPIN-ROTATION SPLITTING BY THE AMOUNT $\gamma_v\left(N + \frac{1}{2}\right)$, WHICH GROWS AS N INCREASES. *Hint*: CONSTRUCT THE 2×2 SECULAR DETERMINANT, USING THE $\left|N\,\frac{1}{2}\,N+\frac{1}{2}\,M\right\rangle$ AND $\left|N\,\frac{1}{2}\,N-\frac{1}{2}\,M\right\rangle$ BASIS FUNCTIONS. USE THE RELATION $2\hat{\mathbf{N}}\cdot\hat{\mathbf{S}} = \hat{\mathbf{J}}^2 - \hat{\mathbf{N}}^2 - \hat{\mathbf{S}}^2$ TO EVALUATE THE MATRIX ELEMENTS OF $\hat{\mathbf{N}}\cdot\hat{\mathbf{S}}$.

The total Hamiltonian for a nonrigid 3D rotor with nonzero electronic spin is

$$\hat{\mathcal{H}} = \hat{\mathcal{H}}_0 + B_v\,\hat{\mathbf{N}}^2 - D_v\,\hat{\mathbf{N}}^4 + \gamma_v\,\hat{\mathbf{N}}\cdot\hat{\mathbf{S}}. \tag{2.35}$$

Defining the basis set

$$\psi_1 = |v\rangle\,\left|N\,\frac{1}{2}\,N+\frac{1}{2}\,M\right\rangle, \tag{2.36}$$

and

$$\psi_2 = |v\rangle\,\left|N\,\frac{1}{2}\,N-\frac{1}{2}\,M\right\rangle, \tag{2.37}$$

we can obtain the energy levels by solving the secular determinant,

$$\begin{vmatrix} \left\langle\psi_1\left|\hat{\mathcal{H}}\right|\psi_1\right\rangle - E & \left\langle\psi_1\left|\hat{\mathcal{H}}\right|\psi_2\right\rangle \\[2mm] \left\langle\psi_2\left|\hat{\mathcal{H}}\right|\psi_1\right\rangle & \left\langle\psi_2\left|\hat{\mathcal{H}}\right|\psi_2\right\rangle - E \end{vmatrix} = 0. \tag{2.38}$$

Because we know that

$$\hat{\mathcal{H}}_0\,|v\rangle = (v + \frac{1}{2})\nu_0\,|v\rangle, \tag{2.39}$$

we need to deal with only the angular part of the basis functions, and we can add the vibrational energy to the eigenvalues at the end. The matrix elements that we must calculate have the form

$$\mathcal{H}_{ij} = \left\langle NSJ_iM_i\left|B_v\,\hat{\mathbf{N}}^2 - D_v\,\hat{\mathbf{N}}^4 + \frac{1}{2}\gamma_v\left(\hat{\mathbf{J}}^2 - \hat{\mathbf{N}}^2 - \hat{\mathbf{S}}^2\right)\right|NSJ_jM_j\right\rangle, \tag{2.40}$$

where we used the relation

$$2\hat{\mathbf{N}}\cdot\hat{\mathbf{S}} = \hat{\mathbf{J}}^2 - \hat{\mathbf{N}}^2 - \hat{\mathbf{S}}^2. \tag{2.41}$$

The diagonal matrix elements are

$$\mathcal{H}_{11} = \left\langle N\,\frac{1}{2}\,N+\frac{1}{2}\,M\,\middle|\,B_v\,\hat{\mathbf{N}}^2 + \frac{1}{2}\gamma_v\left(\hat{\mathbf{J}}^2 - \hat{\mathbf{N}}^2 - \hat{\mathbf{S}}^2\right) - D_v\,\hat{\mathbf{N}}^4\,\middle|\,N\,\frac{1}{2}\,N+\frac{1}{2}\,M\,\right\rangle$$

$$= B_vN(N+1) - D_vN^2(N+1)^2 + \frac{1}{2}\gamma_v\left[J(J+1) - N(N+1) - S(S+1)\right]$$

$$= B_vN(N+1) - D_vN^2(N+1)^2 + \frac{1}{2}\gamma_v\left[(N+\frac{1}{2})(N+\frac{3}{2}) - N(N+1) - \frac{3}{4}\right]$$

$$= B_vN(N+1) - D_vN^2(N+1)^2 + \frac{1}{2}\gamma_vN \tag{2.42}$$

for $J = N + \frac{1}{2}$, and

$$\mathcal{H}_{22} = \left\langle N\,\frac{1}{2}\,N-\frac{1}{2}\,M\,\middle|\,B_v\,\hat{\mathbf{N}}^2 + \frac{1}{2}\gamma_v\left(\hat{\mathbf{J}}^2 - \hat{\mathbf{N}}^2 - \hat{\mathbf{S}}^2\right) - D_v\,\hat{\mathbf{N}}^4\,\middle|\,N\,\frac{1}{2}\,N-\frac{1}{2}\,M\,\right\rangle$$

$$= B_vN(N+1) - D_vN^2(N+1)^2 + \frac{1}{2}\gamma_v\left[(N-\frac{1}{2})(N+\frac{1}{2}) - N(N+1) - \frac{3}{4}\right]$$

$$= B_vN(N+1) - D_vN^2(N+1)^2 - \frac{1}{2}\gamma_v(N+1) \tag{2.43}$$

for $J = N - \frac{1}{2}$. Because ψ_1 and ψ_2 are orthogonal eigenfunctions of $\hat{\mathcal{H}}$, the off-diagonal matrix elements are zero. Substituting equations 2.42 and 2.43 into the determinant gives

$$E_1 = (v+\frac{1}{2})\nu_0 + \mathcal{H}_{11}, \tag{2.44}$$

and

$$E_2 = (v+\frac{1}{2})\nu_0 + \mathcal{H}_{22}. \tag{2.45}$$

For a particular value of N, we obtain a spin-rotation splitting of

$$\Delta E = E_1 - E_2$$

$$= \frac{1}{2}\gamma_vN + \frac{1}{2}\gamma_v(N+1)$$

$$= \gamma_v(N+\frac{1}{2}). \tag{2.46}$$

B.

REPEAT PART A BUT THIS TIME USE THE UNCOUPLED REPRESENTATION $|N\ M_N,\ S\ M_S\rangle$. *Hint:* WORK SPECIFICALLY WITH $\left|N\ M_N,\ \frac{1}{2}\ \frac{1}{2}\right\rangle$ AND $\left|N\ M_N+1,\ \frac{1}{2}\ -\frac{1}{2}\right\rangle$. USE THE IDENTITY

$$\hat{\mathbf{N}}\cdot\hat{\mathbf{S}}\ =\ \hat{N}_x\hat{S}_x+\hat{N}_y\hat{S}_y+\hat{N}_z\hat{S}_z$$

$$=\ \frac{1}{2}(\hat{N}_+\hat{S}_-+\hat{N}_-\hat{S}_+)+\hat{N}_z\hat{S}_z \tag{2.47}$$

TO EVALUATE THE MATRIX ELEMENTS OF $\hat{\mathbf{N}}\cdot\hat{\mathbf{S}}$ APPEARING IN THE 2×2 SECULAR DETERMINANT WITH THESE BASIS FUNCTIONS.

Repeating the evaluation done in part A with the new basis set, we obtain

$$\hat{\mathcal{H}}|N\ M_N,S\ M_S\rangle\ =\ \left[(v+\frac{1}{2})\nu_0+B_vN(N+1)-D_vN^2(N+1)^2+\gamma_v\ \hat{\mathbf{N}}\cdot\hat{\mathbf{S}}\,|N\ M_N,S\ M_S\rangle\right]$$

$$=\ \left[(v+\frac{1}{2})\nu_0+B_vN(N+1)-D_vN^2(N+1)^2\right]|N\ M_N,S\ M_S\rangle$$

$$+\ \gamma_v\left[\frac{1}{2}(\hat{N}_+\hat{S}_-+\hat{N}_-\hat{S}_+)+\hat{N}_z\hat{S}_z\right]|N\ M_N,S\ M_S\rangle\,. \tag{2.48}$$

Writing out the last term for $\left|N\ M_N,S\ \frac{1}{2}\right\rangle$ and $\left|N\ M_N,S\ -\frac{1}{2}\right\rangle$, we obtain

$$\left[\frac{1}{2}(\hat{N}_+\hat{S}_-+\hat{N}_-\hat{S}_+)+\hat{N}_z\hat{S}_z\right]\left|N\ M_N,S\ \frac{1}{2}\right\rangle$$

$$=\ \frac{1}{2}\left(\left[N(N+1)-M_N(M_N+1)\right]\left[S(S+1)-\frac{1}{2}(\frac{1}{2}-1)\right]\right)^{\frac{1}{2}}$$

$$\times\left|N\ M_N+1,S\ -\frac{1}{2}\right\rangle+(M_N)\frac{1}{2}\left|N\ M_N,S\ \frac{1}{2}\right\rangle, \tag{2.49}$$

and

$$\left[\frac{1}{2}(\hat{N}_+\hat{S}_-+\hat{N}_-\hat{S}_+)+\hat{N}_z\hat{S}_z\right]\left|N\ M_N+1,S\ -\frac{1}{2}\right\rangle$$

$$=\ \frac{1}{2}\left(\left[N(N+1)-(M_N+1)M_N\right]\left[S(S+1)+\frac{1}{2}(-\frac{1}{2}+1)\right]\right)^{\frac{1}{2}}$$

$$\times \left| N\ M_N, S\ \frac{1}{2} \right\rangle + (M_N + 1)(-\frac{1}{2}) \left| N\ M_N + 1, S\ -\frac{1}{2} \right\rangle, \qquad (2.50)$$

where $S = \frac{1}{2}$. Here we have used

$$\hat{S}_+ \left| S\ \frac{1}{2} \right\rangle = 0 \quad \text{and} \quad \hat{S}_- \left| S\ -\frac{1}{2} \right\rangle = 0. \qquad (2.51)$$

Now we can evaluate each of the matrix elements $\mathcal{H}_{ij}$, noting that $\hat{N}_+\hat{S}_-$ and $\hat{N}_-\hat{S}_+$ do not contribute to the diagonal terms and that $\hat{N}_z\hat{S}_z$ does not contribute to the off-diagonal terms:

$$\mathcal{H}_{11} = \left[(v + \frac{1}{2})\nu_0 + B_v N(N+1) - D_v N^2(N+1)^2 \right] \left\langle N\ M_N, S\ \frac{1}{2} \middle| N\ M_N, S\ \frac{1}{2} \right\rangle$$

$$+ \frac{\gamma_v}{2} \left(\left[N(N+1) - M_N(M_N+1) \right] \left[S(S+1) - \frac{1}{2}(\frac{1}{2} - 1) \right] \right)^{\frac{1}{2}}$$

$$\times \left\langle N\ M_N, S\ \frac{1}{2} \middle| N\ M_N + 1, S\ -\frac{1}{2} \right\rangle + \gamma_v \frac{1}{2} M_N \left\langle N\ M_N, S\ \frac{1}{2} \middle| N\ M_N, S\ \frac{1}{2} \right\rangle$$

$$= (v + \frac{1}{2})\nu_0 + B_v N(N+1) - D_v N^2(N+1)^2 + \frac{1}{2}\gamma_v M_N, \qquad (2.52)$$

$$\mathcal{H}_{22} = \left[(v + \frac{1}{2})\nu_0 + B_v N(N+1) - D_v N^2(N+1)^2 - \gamma_v \frac{1}{2}(M_N + 1) \right]$$

$$\times \left\langle N\ M_N + 1, S\ -\frac{1}{2} \middle| N\ M_N + 1, S\ -\frac{1}{2} \right\rangle$$

$$+ \frac{\gamma_v}{2} \left(\left[N(N+1) - (M_N+1)M_N \right] \left[S(S+1) + \frac{1}{2}(-\frac{1}{2} + 1) \right] \right)^{\frac{1}{2}}$$

$$\times \left\langle N\ M_N + 1, S\ -\frac{1}{2} \middle| N\ M_N, S\ \frac{1}{2} \right\rangle$$

$$= (v + \frac{1}{2})\nu_0 + B_v N(N+1) - D_v N^2(N+1)^2 - \frac{1}{2}\gamma_v(M_N + 1), \qquad (2.53)$$

$$\mathcal{H}_{12} = \left[(v + \frac{1}{2})\nu_0 + B_v N(N+1) - D_v N^2(N+1)^2 \right] \left\langle N\ M_N, S\ \frac{1}{2} \middle| N\ M_N + 1, S\ -\frac{1}{2} \right\rangle$$

$$+ \frac{\gamma_v}{2} \left(\left[N(N+1) - (M_N+1)M_N \right] \left[\frac{1}{2}(\frac{1}{2}+1) + \frac{1}{2}(-\frac{1}{2}+1) \right] \right)^{\frac{1}{2}}$$

$$\times \left\langle N \ M_N, S \ \frac{1}{2} \middle| N \ M_N, S \ \frac{1}{2} \right\rangle + \gamma_v(-\frac{1}{2})(M_N+1) \left\langle N \ M_N, S \ \frac{1}{2} \middle| N \ M_N+1, S \ -\frac{1}{2} \right\rangle$$

$$= \frac{1}{2}\gamma_v[N(N+1) - (M_N+1)M_N]^{\frac{1}{2}}, \tag{2.54}$$

and

$$\mathcal{H}_{21} = \mathcal{H}_{12}. \tag{2.55}$$

In evaluating $\mathcal{H}_{ij}$, we have used the orthonormality of the wave functions and the real Hermitian property of $\hat{\mathcal{H}}$. Having all the $\mathcal{H}_{ij}$, we can solve the secular equation,

$$(\mathcal{H}_{11} - E)(\mathcal{H}_{22} - E) - \mathcal{H}_{12}^2 = 0.$$

Let us define β as

$$\beta = (v + \frac{1}{2})\nu_0 + B_v N(N+1) - D_v N^2(N+1)^2 - E. \tag{2.56}$$

The secular equation becomes

$$0 = (\beta + \frac{1}{2}\gamma_v M_N)[\beta - \frac{1}{2}\gamma_v(M_N+1)] - \frac{\gamma_v^2}{4}[N(N+1) - (M_N+1)M_N]$$

$$= \beta^2 + \beta[\frac{1}{2}\gamma_v M_N - \frac{1}{2}\gamma_v(M_N+1)] - \frac{\gamma_v^2}{4}[M_N(M_N+1) + N(N+1) - (M_N+1)M_N]$$

$$= \beta^2 - \frac{1}{2}\gamma_v\beta - \frac{\gamma_v^2}{4}N(N+1)$$

$$= \left[\beta + \frac{1}{2}\gamma_v N\right] \left[\beta - \frac{1}{2}\gamma_v(N+1)\right], \tag{2.57}$$

from which we get the two solutions

$$\beta = -\frac{1}{2}\gamma_v N \tag{2.58}$$

and

$$\beta = \frac{1}{2}\gamma_v(N+1). \tag{2.59}$$

Substituting equations 2.58 and 2.59 into equation 2.56, we obtain the energies

$$E_1 = (v + \frac{1}{2})\nu_0 + B_v N(N+1) - D_v N^2(N+1)^2 + \frac{1}{2}\gamma_v N \tag{2.60}$$

and

$$E_2 = (v + \frac{1}{2})\nu_0 + B_v N(N + 1) - D_v N^2(N + 1)^2 - \frac{1}{2}\gamma_v(N + 1). \tag{2.61}$$

As in part A, we finally obtain

$$\Delta E = \gamma_v(N + \frac{1}{2}). \tag{2.62}$$

C.

EVALUATE $\Delta E_z(M)$ BY REWRITING $|NSJM\rangle$ IN TERMS OF $|N\ M_N, S\ M_S\rangle$ AND THEN CALCULATING THE MATRIX ELEMENTS OF $\hat{\mathcal{H}}_z$. SHOW THAT

$$\Delta E_z(M) = \pm g_s\mu_0 H\frac{M}{2N + 1}, \tag{2.63}$$

WHERE THE UPPER SIGN IS FOR $J = N + \frac{1}{2}$ AND THE LOWER SIGN IS FOR $J = N - \frac{1}{2}$.

We know that in general

$$|JM\rangle = |NSJM\rangle$$

$$= \sum_{M_N, M_S} |N\ M_N, S\ M_S\rangle \langle N\ M_N, S\ M_S\ |NSJM\rangle. \tag{2.64}$$

For $J = N + \frac{1}{2}$ we have

$$\left|N + \frac{1}{2}\ M\right\rangle = \left|N\ M_N, \frac{1}{2}\ -\frac{1}{2}\right\rangle \left\langle N\ M_N, \frac{1}{2}\ -\frac{1}{2}\right|N + \frac{1}{2}\ M\right\rangle$$

$$+ \left|N\ M_N, \frac{1}{2}\ \frac{1}{2}\right\rangle \left\langle N\ M_N, \frac{1}{2}\ \frac{1}{2}\right|N + \frac{1}{2}\ M\right\rangle$$

$$= C_{-\frac{1}{2}}\left|N\ M_N, \frac{1}{2}\ -\frac{1}{2}\right\rangle + C_{\frac{1}{2}}\left|N\ M_N, \frac{1}{2}\ \frac{1}{2}\right\rangle, \tag{2.65}$$

where $C_{\pm\frac{1}{2}}$ are Clebsch-Gordan coefficients. For the Zeeman effect in $^2\Sigma$ molecules, we can evaluate the energy of each magnetic sublevel M of level J by first-order perturbation theory:

$$\Delta E_z(M) = \left\langle NSJM\left|\hat{\mathcal{H}}_z\right|NSJM\right\rangle, \tag{2.66}$$

where

$$\hat{\mathcal{H}}_z = g_s\mu_0 H\hat{S}_z. \tag{2.67}$$

Substituting equation 2.65 into equation 2.66,

$$\Delta E_z(M) = \left|C_{-\frac{1}{2}}\right|^2 \left\langle N\ M_N, \frac{1}{2}\ -\frac{1}{2} \middle| g_s\mu_0 H \hat{S}_z \middle| N\ M_N, \frac{1}{2}\ -\frac{1}{2} \right\rangle$$

$$+ \left|C_{\frac{1}{2}}\right|^2 \left\langle N\ M_N, \frac{1}{2}\ \frac{1}{2} \middle| g_s\mu_0 H \hat{S}_z \middle| N\ M_N, \frac{1}{2}\ \frac{1}{2} \right\rangle$$

$$+ C_{-\frac{1}{2}}^* C_{\frac{1}{2}} \left\langle N\ M_N, \frac{1}{2}\ -\frac{1}{2} \middle| g_s\mu_0 H \hat{S}_z \middle| N\ M_N, \frac{1}{2}\ \frac{1}{2} \right\rangle$$

$$+ C_{\frac{1}{2}}^* C_{-\frac{1}{2}} \left\langle N\ M_N, \frac{1}{2}\ \frac{1}{2} \middle| g_s\mu_0 H \hat{S}_z \middle| N\ M_N, \frac{1}{2}\ -\frac{1}{2} \right\rangle$$

$$= \left|C_{-\frac{1}{2}}\right|^2 g_s\mu_0 H\left(-\frac{1}{2}\right) + \left|C_{\frac{1}{2}}\right|^2 g_s\mu_0 H\left(\frac{1}{2}\right)$$

$$= \frac{1}{2} g_s\mu_0 H \left(\left|C_{\frac{1}{2}}\right|^2 - \left|C_{-\frac{1}{2}}\right|^2 \right). \tag{2.68}$$

To get the final answer, we evaluate $C_{\pm\frac{1}{2}}$ from Table Z-2.4B,

$$C_{-\frac{1}{2}} = \left\langle N\ M+\frac{1}{2}, \frac{1}{2}\ -\frac{1}{2}\ \middle|\ N+\frac{1}{2}\ M \right\rangle$$

$$= \sqrt{\frac{N-M+\frac{1}{2}}{2N+1}} \tag{2.69}$$

$$C_{\frac{1}{2}} = \left\langle N\ M-\frac{1}{2}, \frac{1}{2}\ \frac{1}{2}\ \middle|\ N+\frac{1}{2}\ M \right\rangle$$

$$= \sqrt{\frac{N+M+\frac{1}{2}}{2N+1}}, \tag{2.70}$$

and then substitute into equation 2.68 to obtain

$$\Delta E_z = \frac{1}{2} g_s\mu_0 H \left(\frac{N+M+\frac{1}{2}}{2N+1} - \frac{N-M+\frac{1}{2}}{2N+1} \right)$$

$$= \frac{1}{2} g_s\mu_0 H\ \frac{2M}{2N+1}$$

$$= g_s\mu_0 H \frac{M}{2N+1}. \tag{2.71}$$

For $J = N - \frac{1}{2}$, we have

$$\left|N - \frac{1}{2} M\right\rangle = \left|N\, M_N, \frac{1}{2}\, -\frac{1}{2}\right\rangle \left\langle N\, M_N, \frac{1}{2}\, -\frac{1}{2}\,\right| N - \frac{1}{2} M\right\rangle$$

$$+ \left|N\, M_N, \frac{1}{2}\, \frac{1}{2}\right\rangle \left\langle N\, M_N, \frac{1}{2}\, \frac{1}{2}\,\right| N - \frac{1}{2} M\right\rangle$$

$$= D_{-\frac{1}{2}} \left|N\, M_N, \frac{1}{2}\, -\frac{1}{2}\right\rangle + D_{\frac{1}{2}} \left|N\, M_N, \frac{1}{2}\, \frac{1}{2}\right\rangle, \tag{2.72}$$

where $D_{\pm\frac{1}{2}}$ are the corresponding Clebsch-Gordon coefficients. Substituting this result into equation 2.66, we get

$$\Delta E_z(M) = g_s\mu_0 H \frac{1}{2}\left(\left|D_{\frac{1}{2}}\right|^2 - \left|D_{-\frac{1}{2}}\right|^2\right)$$

$$= g_s\mu_0 H \frac{1}{2}\left(\frac{N - M + \frac{1}{2}}{2N+1} - \frac{N + M + \frac{1}{2}}{2N+1}\right)$$

$$= -g_s\mu_0 H \frac{M}{2N+1}. \tag{2.73}$$

D.

WORK IN THE COUPLED REPRESENTATION TO OBTAIN AN EXPLICIT EXPRESSION FOR THE ENERGIES OF THE MAGNETIC SUBLEVELS OF A $^2\Sigma$ MOLECULE IN A MAGNETIC FIELD H. SHOW THAT FOR SMALL FIELD STRENGTHS THESE EXPRESSIONS APPROACH THE RESULTS OF FIRST-ORDER PERTURBATION THEORY, IN WHICH THE ENERGIES OF THE MAGNETIC SUBLEVELS VARY LINEARLY WITH H.

To calculate the energies of the magnetic sublevels, we have to solve the 2×2 secular determinant, where the total Hamiltonian is now

$$\hat{\mathcal{H}} = \hat{\mathcal{H}}_0 + B_v\, \hat{\mathbf{N}}^2 + \gamma_v\, \hat{\mathbf{N}} \cdot \hat{\mathbf{S}} - D_v\, \hat{\mathbf{N}}^4 + g_s\mu_0 H \hat{\mathbf{S}}_z. \tag{2.74}$$

Using the answers of parts A and C, we can easily calculate the matrix elements $\mathcal{H}_{ij}$. For example,

$$\mathcal{H}_{11} = \left\langle N\, \frac{1}{2}\, N+\frac{1}{2}\, M\right| \hat{\mathcal{H}}_0 + B_v\, \hat{\mathbf{N}}^2 + \gamma_v\, \hat{\mathbf{N}} \cdot \hat{\mathbf{S}} - D_v\, \hat{\mathbf{N}}^4 + g_s\mu_0 H \hat{\mathbf{S}}_z \left| N\, \frac{1}{2}\, N+\frac{1}{2}\, M\right\rangle$$

$$= (v + \frac{1}{2})\nu_0 + B_v N(N+1) - D_v N^2(N+1)^2 + \frac{1}{2}\gamma_v N + g_s\mu_0 H \frac{M}{2N+1}, \tag{2.75}$$

where we used equations 2.42 and 2.71.

In the same way, we obtain

$$\mathcal{H}_{22} = \left\langle N\,\frac{1}{2}\,N-\frac{1}{2}\,M\,\middle|\,\hat{\mathcal{H}}\,\middle|\,N\,\frac{1}{2}\,N-\frac{1}{2}\,M\right\rangle$$

$$= (v+\frac{1}{2})\nu_0 + B_v N(N+1) - D_v N^2(N+1)^2 - \frac{1}{2}\gamma_v(N+1) - g_s\mu_0 H\frac{M}{2N+1}, \tag{2.76}$$

using in this case equations 2.43 and 2.73. Finally, using equations 2.65, 2.72, and 2.54, we obtain

$$\mathcal{H}_{12} = \left\langle N\,\frac{1}{2}\,N+\frac{1}{2}\,M\,\middle|\,\hat{\mathcal{H}}\,\middle|\,N\,\frac{1}{2}\,N-\frac{1}{2}\,M\right\rangle$$

$$= g_s\mu_0 H\left\langle N\,\frac{1}{2}\,N+\frac{1}{2}\,M\,\middle|\,\hat{\mathbf{S}}_z\,\middle|\,N\,\frac{1}{2}\,N-\frac{1}{2}\,M\right\rangle$$

$$= g_s\mu_0 H\left[C_{-\frac{1}{2}}D_{-\frac{1}{2}}\left\langle N\,M_N,\frac{1}{2}\,-\frac{1}{2}\,\middle|\,\hat{\mathbf{S}}_z\,\middle|\,N\,M_N,\frac{1}{2}\,-\frac{1}{2}\right\rangle\right.$$

$$\left. + C_{\frac{1}{2}}D_{\frac{1}{2}}\left\langle N\,M_N,\frac{1}{2}\,\frac{1}{2}\,\middle|\,\hat{\mathbf{S}}_z\,\middle|\,N\,M_N,\frac{1}{2}\,\frac{1}{2}\right\rangle\right]$$

$$= -\frac{1}{2}g_s\mu_0 H\left[C_{-\frac{1}{2}}D_{-\frac{1}{2}} - C_{\frac{1}{2}}D_{\frac{1}{2}}\right]$$

$$= -\frac{1}{2}g_s\mu_0 H\left[\left(\frac{N-M+\frac{1}{2}}{2N+1}\right)^{\frac{1}{2}}\left(\frac{N+M+\frac{1}{2}}{2N+1}\right)^{\frac{1}{2}} + \left(\frac{N+M+\frac{1}{2}}{2N+1}\right)^{\frac{1}{2}}\left(\frac{N-M+\frac{1}{2}}{2N+1}\right)^{\frac{1}{2}}\right]$$

$$= -\frac{g_s\mu_0 H}{2N+1}\sqrt{(N+\frac{1}{2})^2 - M^2}. \tag{2.77}$$

Again, we can say that

$$\mathcal{H}_{12} = \mathcal{H}_{21} \tag{2.78}$$

because the Hamiltonian is Hermitian and $\mathcal{H}_{12}$ is real.

Having evaluated all the matrix elements, we can now solve the secular equation

$$(\mathcal{H}_{11} - E)(\mathcal{H}_{22} - E) - \mathcal{H}_{12}^2 = 0. \tag{2.79}$$

Defining

$$\beta = (v+\frac{1}{2})\nu_0 + B_v N(N+1) - D_v N^2(N+1)^2 - E, \tag{2.80}$$

we get

$$0 = \left[\beta + \frac{1}{2}\gamma_v N + g_s\mu_0 H\frac{M}{2N+1}\right]\left[\beta - \frac{1}{2}\gamma_v(N+1)\right.$$

$$\left. -g_s\mu_0 H\frac{M}{2N+1}\right] - \left(\frac{g_s\mu_0 H}{2N+1}\right)^2\left[(N+\frac{1}{2})^2 - M^2\right]$$

$$= \beta^2 + \beta\left[\frac{1}{2}\gamma_v N + g_s\mu_0 H\frac{M}{2N+1} - \frac{1}{2}\gamma_v(N+1) - g_s\mu_0 H\frac{M}{2N+1}\right]$$

$$-\left(\frac{1}{2}\gamma_v N + g_s\mu_0 H\frac{M}{2N+1}\right)\left[\frac{1}{2}\gamma_v(N+1) + g_s\mu_0 H\frac{M}{2N+1}\right]$$

$$-\left(\frac{g_s\mu_0 H}{2N+1}\right)^2\left[(N+\frac{1}{2})^2 - M^2\right]$$

$$= \beta^2 - \beta\frac{\gamma_v}{2} - \frac{\gamma_v^2}{4}N(N+1) - \frac{\gamma_v}{2}g_s\mu_0 HM - \left(\frac{g_s\mu_0 H}{2}\right)^2. \tag{2.81}$$

It follows from equation 2.81 that

$$\beta = \frac{1}{2}\left\{\frac{\gamma_v}{2} \pm \left[\left(\frac{\gamma_v}{2}\right)^2 + \gamma_v^2 N(N+1) + 2\gamma_v g_s\mu_0 HM + (g_s\mu_0 H)^2\right]^{\frac{1}{2}}\right\}$$

$$= \frac{\gamma_v}{4} \pm \frac{1}{2}\left[\gamma_v^2(N+\frac{1}{2})^2 + g_s\mu_0 H(2\gamma_v M + g_s\mu_0 H)\right]^{\frac{1}{2}}. \tag{2.82}$$

Substituting equation 2.80 into 2.82 we obtain

$$E = (v+\frac{1}{2})\nu_0 + B_v N(N+1) - D_v N^2(N+1)^2 - \frac{\gamma_v}{4}$$

$$\pm\frac{1}{2}\left[\gamma_v^2(N+\frac{1}{2})^2 + g_s\mu_0 H(2\gamma_v M + g_s\mu_0 H)\right]^{\frac{1}{2}}. \tag{2.83}$$

The next step is to analyze what happens in the case of a weak field. Equation 2.83 can be rewritten in the form

$$E = (v+\frac{1}{2})\nu_0 + B_v N(N+1) - D_v N^2(N+1)^2 - \frac{\gamma_v}{4}$$

$$\pm \frac{\gamma_v}{4}(2N+1)\left[1+\frac{4g_s\mu_0 H}{\gamma_v^2(2N+1)^2}(2\gamma_v M+g_s\mu_0 H)\right]^{\frac{1}{2}}.$$

$$(2.84)$$

We define the dimensionless quantity

$$x=\frac{4g_s\mu_0 H}{\gamma_v^2(2N+1)^2}(2\gamma_v M+g_s\mu_0 H). \qquad (2.85)$$

In the weak field limit $x \to 0$, and a Taylor series expansion gives

$$(1+x)^{\frac{1}{2}} \approx 1+\frac{1}{2}x+\dots . \qquad (2.86)$$

Equation 2.83 becomes

$$E \quad \approx \quad (v+\frac{1}{2})\nu_0+B_v N(N+1)-D_v N^2(N+1)^2-\frac{\gamma_v}{4}$$

$$\pm\frac{\gamma_v}{4}(2N+1)\left[1+\frac{2g_s\mu_0 H}{\gamma_v^2(2N+1)^2}(2\gamma_v M+g_s\mu_0 H)\right]$$

$$= \quad (v+\frac{1}{2})\nu_0+B_v N(N+1)-D_v N^2(N+1)^2-\frac{\gamma_v}{4}$$

$$\pm\left[\frac{\gamma_v}{4}(2N+1)+g_s\mu_0 H\frac{M}{2N+1}+\frac{(g_s\mu_0 H)^2}{2\gamma_v(2N+1)}\right]. \qquad (2.87)$$

When only weak fields are present we may ignore the quadratic term in H. We therefore obtain

$$E_1 \approx (v+\frac{1}{2})\nu_0+B_v N(N+1)-D_v N^2(N+1)^2+\frac{\gamma_v}{2}N+g_s\mu_0 H\frac{M}{2N+1}, \qquad (2.88)$$

and

$$E_2 \approx (v+\frac{1}{2})\nu_0+B_v N(N+1)-D_v N^2(N+1)^2-\frac{\gamma_v}{2}(N+1)-g_s\mu_0 H\frac{M}{2N+1}, \qquad (2.89)$$

which is the same result that we obtained from first-order perturbation theory.

E.

PLOT A GRAPH SHOWING THE VARIATION OF THE $^2\Sigma$ ENERGY LEVELS WITH MAGNETIC FIELD STRENGTH FOR THE LOWEST EIGHT LEVELS THAT CAN BE CONSTRUCTED FROM $N=0, S=\frac{1}{2}$, AND $N=1, S=\frac{1}{2}$. USE $B_v=1$ cm^{-1}, $D_v=1\times 10^{-6}$ cm^{-1}, AND $\gamma_v=0.01$ cm^{-1}.

Consider the eight states in both representations:

State	N	$\|NSJM\rangle$ coupled	$\|NM_N, SM_S\rangle$ uncoupled
A	0	$\|0\frac{1}{2}\frac{1}{2}\,\frac{1}{2}\rangle$	$\|00,\frac{1}{2}\,\frac{1}{2}\rangle$
B	0	$\|0\frac{1}{2}\frac{1}{2}-\frac{1}{2}\rangle$	$\|00,\frac{1}{2}-\frac{1}{2}\rangle$
C	1	$\|1\frac{1}{2}\frac{3}{2}\,\frac{3}{2}\rangle$	$\|11,\frac{1}{2}\,\frac{1}{2}\rangle$
D	1	$\|1\frac{1}{2}\frac{3}{2}-\frac{3}{2}\rangle$	$\|1-1,\frac{1}{2}-\frac{1}{2}\rangle$
E	1	$\|1\frac{1}{2}\frac{3}{2}\,\frac{1}{2}\rangle$	$\sqrt{\frac{1}{3}}\,\|11,\frac{1}{2}-\frac{1}{2}\rangle+\sqrt{\frac{2}{3}}\,\|1\,0,\frac{1}{2}\frac{1}{2}\rangle$
F	1	$\|1\frac{1}{2}\frac{1}{2}\,\frac{1}{2}\rangle$	$\sqrt{\frac{2}{3}}\,\|11,\frac{1}{2}-\frac{1}{2}\rangle-\sqrt{\frac{1}{3}}\,\|1\,0,\frac{1}{2}\frac{1}{2}\rangle$
G	1	$\|1\frac{1}{2}\frac{3}{2}-\frac{1}{2}\rangle$	$\sqrt{\frac{2}{3}}\,\|10,\frac{1}{2}-\frac{1}{2}\rangle+\sqrt{\frac{1}{3}}\,\|1-1,\frac{1}{2}\frac{1}{2}\rangle$
H	1	$\|1\frac{1}{2}\frac{1}{2}-\frac{1}{2}\rangle$	$\sqrt{\frac{1}{3}}\,\|11,\frac{1}{2}-\frac{1}{2}\rangle-\sqrt{\frac{2}{3}}\,\|1-1,\frac{1}{2}\frac{1}{2}\rangle$

In the uncoupled representation, states E and F and states G and H mix with each other. For the states that do not mix (A, B, C, and D), the off-diagonal matrix elements are zero and the total energy can be written as the sum of two diagonal contributions, the non-Zeeman term calculated in part A and the Zeeman term calculated in part C. For the mixed states, we have solved the secular equation in part D. We can also see that the energies of pairs of states (A,B), (C,D), (E,G), and (F,H) have common limiting values when $H = 0$.

The energies of the uncoupled states are plotted in Figure 2.1, and the energies of the coupled states are plotted in Figure 2.2.

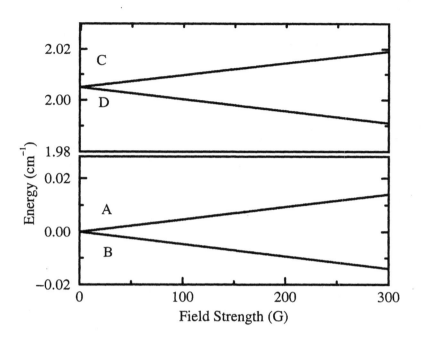

Figure 2.1: Energies of the uncoupled states, A, B, C, and D.

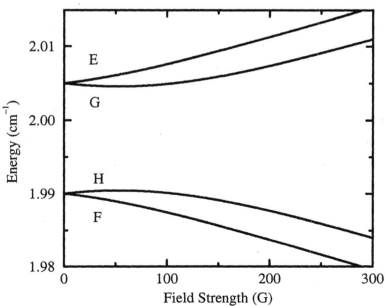

Figure 2.2: Energies of the coupled states, E, F, G, and H.

Chapter 3

TRANSFORMATION UNDER ROTATION

APPLICATION 4
ENERGY LEVELS OF ATOMS
WITH TWO VALENCE ELECTRONS

A.

SHOW THAT l_{iz} AND l_{jz} DO *not* COMMUTE WITH THE NONRELATIVISTIC HAMILTONIAN BY FINDING EXPLICITLY THE NONZERO VALUES OF THE COMMUTATORS

$$\left[l_{iz}, \frac{e^2}{r_{ij}} \right] \quad \text{AND} \quad \left[l_{jz}, \frac{e^2}{r_{ij}} \right].$$

Operating on an arbitrary wave function ϕ,

$$\left[l_{iz}, \frac{e^2}{r_{ij}} \right] \phi = e^2 \left(l_{iz} \frac{1}{r_{ij}} - \frac{1}{r_{ij}} l_{iz} \right) \phi$$

$$= e^2 \left[-i \left(x_i \frac{\partial}{\partial y_i} - y_i \frac{\partial}{\partial x_i} \right) \frac{1}{r_{ij}} + i \frac{1}{r_{ij}} \left(x_i \frac{\partial}{\partial y_i} - y_i \frac{\partial}{\partial x_i} \right) \right] \phi$$

$$= -ie^2 \left(x_i \frac{\partial r_{ij}^{-1}}{\partial y_i} + r_{ij}^{-1} x_i \frac{\partial}{\partial y_i} - y_i \frac{\partial r_{ij}^{-1}}{\partial x_i} - y_i r_{ij}^{-1} \frac{\partial}{\partial x_i} - r_{ij}^{-1} x_i \frac{\partial}{\partial y_i} + r_{ij}^{-1} y_i \frac{\partial}{\partial x_i} \right) \phi$$

$$= -ie^2 \left(x_i \frac{\partial r_{ij}^{-1}}{\partial y_i} - y_i \frac{\partial r_{ij}^{-1}}{\partial x_i} \right) \phi$$

$$= \frac{-ie^2}{r_{ij}^3} \left[x_i (y_j - y_i) - y_i (x_j - x_i) \right] \phi$$

$$= \frac{-ie^2}{r_{ij}^3} (x_i y_j - x_j y_i) \phi$$

$$= -ie^2 \frac{(\vec{r_i} \times \vec{r_j})_z}{r_{ij}^3} \phi, \tag{3.1}$$

where we have used the identity,

$$\frac{\partial r_{ij}^{-1}}{\partial q_i} = -\frac{q_i - q_j}{r_{ij}^3}, \tag{3.2}$$

with $q_i = x_i, y_i, z_i$, and $q_j = x_j, y_j, z_j$.

Repeating the derivation for l_{jz}, we get

$$\left[l_{jz}, \frac{e^2}{r_{ij}} \right] \phi = -ie^2 \frac{(\vec{r_j} \times \vec{r_i})_z}{r_{ij}^3} \phi. \tag{3.3}$$

B.

USE THE PRECEDING RESULTS TO SHOW THAT THE SUM $l_{iz} + l_{jz}$ COMMUTES WITH $\frac{e^2}{r_{ij}}$.

In part A, we proved that

$$\left[l_{iz} + l_{jz}, \frac{e^2}{r_{ij}} \right] = \left[l_{iz}, \frac{e^2}{r_{ij}} \right] + \left[l_{jz}, \frac{e^2}{r_{ij}} \right]$$

$$= \frac{-ie^2}{r_{ij}^3} \left[(\vec{r_i} \times \vec{r_j})_z + (\vec{r_j} \times \vec{r_i})_z \right]. \tag{3.4}$$

From the vector property

$$(\vec{r_i} \times \vec{r_j})_z = -(\vec{r_j} \times \vec{r_i})_z, \tag{3.5}$$

equation 3.4 becomes

$$\left[l_{iz} + l_{jz}, \frac{e^2}{r_{ij}} \right] = 0. \tag{3.6}$$

C.

CONSIDER TWO PARTICLES 1 AND 2 HAVING ORBITAL ANGULAR MOMENTA l_1 AND l_2 WITH PROJECTIONS m_{l_1} AND m_{l_2}. WE MAY WRITE THE COUPLED REPRESENTATION IN EITHER OF TWO WAYS:

$$|l_1 \ l_2 \ L \ M_L\rangle = \sum_{m_{l_1} \ m_{l_2}} \langle l_1 \ m_{l_1}, \ l_2 \ m_{l_2} | L \ M_L \rangle |l_1 \ m_{l_1}\rangle |l_2 \ m_{l_2}\rangle \tag{3.7}$$

OR

$$|l_2 \ l_1 \ L \ M_L\rangle = \sum_{m_{l_1} \ m_{l_2}} \langle l_2 \ m_{l_2}, \ l_1 \ m_{l_1} | L \ M_L \rangle |l_1 \ m_{l_1}\rangle |l_2 \ m_{l_2}\rangle. \tag{3.8}$$

SHOW THAT

$$|l_1 \ l_2 \ L \ M_L\rangle = (-1)^{l_1 + l_2 - L} |l_2 \ l_1 \ L \ M_L\rangle. \tag{3.9}$$

From the properties of the Clebsch-Gordan coefficients given in equation Z-2.26, we know that

$$\langle l_1 \ m_{l_1}, \ l_2 \ m_{l_2} | L \ M_L \rangle = (-1)^{l_1 + l_2 - L} \langle l_2 \ m_{l_2}, \ l_1 \ m_{l_1} | L \ M_L \rangle. \tag{3.10}$$

Inserting equation 3.10 into equations 3.7 and 3.8, we obtain

$$|l_1 \ l_2 \ L \ M_L\rangle = (-1)^{l_1 + l_2 - L} |l_2 \ l_1 \ L \ M_L\rangle. \tag{3.11}$$

D.

DETERMINE THE TERMS THAT ARISE FROM THE $(ns)^2$ CONFIGURATION, THE $(np)^2$ CONFIGURATION, AND THE $(nd)^2$ CONFIGURATION.

For the $(ns)^2$ configuration there are two equivalent electrons with $l_1 = l_2 = 0$ and $s_1 = s_2 = \frac{1}{2}$. The possible values of L and S are

$L = 0$

$S = 0, 1.$

For the total wave function to be antisymmetric, it is necessary that $L + S$ be even. The only possible term is therefore ^1S.

For the $(np)^2$ configuration, $l_1 = l_2 = 1$ and $s_1 = s_2 = \frac{1}{2}$. The possible values of L and S are

$L = 0, 1, 2$

$S = 0, 1.$

Because $L + S$ should be even, we obtain the following three terms:

$^1S \quad (L = 0, \ S = 0)$

$^3P \quad (L = 1, \ S = 1)$

$^1D \quad (L = 2, \ S = 0).$

For the $(nd)^2$ configurations, the possible values of L and S are

$L = 0, 1, 2, 3, 4$

$S = 0, 1,$

and the possible terms are

$^1S \quad (L = 0, \ S = 0)$

$^3P \quad (L = 1, \ S = 1)$

$^1D \quad (L = 2, \ S = 0)$

$^3F \quad (L = 3, \ S = 1)$

$^1G \quad (L = 4, \ S = 0).$

We can check these results by counting the number of states in the uncoupled representation. The number of distinct sets of quantum numbers $(m_{l_1}, m_{s_1}, m_{l_2}, m_{s_2})$ for a pair of indistinguishable electrons is

$$N = \frac{1}{2} \left[(2l+1)2\right] \left[(2l+1)2 - 1\right]. \tag{3.12}$$

It is easy to confirm by a direct count of the term symbols derived above that $N[(ns)^2] = 1$, $N[(np)^2] = 15$, and $N[(np)^2] = 45$.

E.

SHOW THAT $\hat{S}^2$ ACTS ON THE ANTISYMMETRIC SPIN COMBINATION TO GIVE $S(S+1) = 0$ AND THE SYM-METRIC COMBINATION TO GIVE $S(S+1) = 2$. COMBINING THESE RESULTS, WE CONCLUDE THAT THE PLUS SIGN IS FOR THE SINGLET STATE (^1L) AND THE MINUS SIGN IS FOR A TRIPLET STATE (^3L).

We can write the operator $\hat{S}^2$ as

$$
\begin{aligned}
\hat{S}^2 &= \hat{S}_1^2 + \hat{S}_2^2 + 2(\hat{S}_{1x}\hat{S}_{2x} + \hat{S}_{1y}\hat{S}_{2y} + \hat{S}_{1z}\hat{S}_{2z}) \\[2mm]
&= \hat{S}_1^2 + \hat{S}_2^2 + 2\hat{S}_{1z}\hat{S}_{2z} + \frac{1}{2}(\hat{S}_{1+} + \hat{S}_{1-})(\hat{S}_{2+} + \hat{S}_{2-}) - \frac{1}{2}(\hat{S}_{1+} - \hat{S}_{1-})(\hat{S}_{2+} - \hat{S}_{2-}) \\[2mm]
&= \hat{S}_1^2 + \hat{S}_2^2 + 2\hat{S}_{1z}\hat{S}_{2z} + \hat{S}_{1+}\hat{S}_{2-} + \hat{S}_{1-}\hat{S}_{2+}.
\end{aligned}
\tag{3.13}
$$

For the antisymmetric combination we have

$$
|\psi_{anti}\rangle = \frac{1}{\sqrt{2}}\left[|\alpha(1)\beta(2)\rangle - |\beta(1)\alpha(2)\rangle\right]
\tag{3.14}
$$

and

$$
\begin{aligned}
\hat{S}^2|\psi_{anti}\rangle &= \frac{1}{\sqrt{2}}\left[\left(\frac{3}{4} + \frac{3}{4}\right)\left[|\alpha(1)\beta(2)\rangle - |\beta(1)\alpha(2)\rangle\right]\right. \\[4mm]
&\quad \left. - \frac{1}{2}\left[|\alpha(1)\beta(2)\rangle - |\beta(1)\alpha(2)\rangle\right] + |\beta(1)\alpha(2)\rangle - |\alpha(1)\beta(2)\rangle\right] \\[4mm]
&= 0.
\end{aligned}
\tag{3.15}
$$

Therefore, we obtain $S(S+1) = 0$, which yields $S = 0$.

For the symmetric combination, the possible functions are

$$
|\psi_{sym_1}\rangle = \frac{1}{\sqrt{2}}\left[|\alpha(1)\beta(2)\rangle + |\alpha(2)\beta(1)\rangle\right],
\tag{3.16}
$$

$$
|\psi_{sym_2}\rangle = |\alpha(1)\alpha(2)\rangle,
\tag{3.17}
$$

and

$$
|\psi_{sym_3}\rangle = |\beta(1)\beta(2)\rangle.
\tag{3.18}
$$

Evaluating the expectation value of $\hat{S}^2$ for each of these functions, we obtain

$$
\begin{aligned}
\hat{S}^2|\psi_{sym_1}\rangle &= \frac{1}{\sqrt{2}}\left[\left(\frac{3}{4} + \frac{3}{4}\right)\left[|\alpha(1)\beta(2)\rangle + |\beta(1)\alpha(2)\rangle\right]\right. \\[4mm]
&\quad \left. - \frac{1}{2}\left[|\alpha(1)\beta(2)\rangle + |\beta(1)\alpha(2)\rangle\right] + |\beta(1)\alpha(2)\rangle + |\alpha(1)\beta(2)\rangle\right]
\end{aligned}
$$

$$= \ 2 |\psi_{sym_1}\rangle, \tag{3.19}$$

$$\hat{S}^2 |\psi_{sym_2}\rangle \ = \ \left(\frac{3}{4}+\frac{3}{4}\right) |\alpha(1)\alpha(2)\rangle + 2 \times \frac{1}{2} \times \frac{1}{2} |\alpha(1)\alpha(2)\rangle$$

$$= \ 2 |\psi_{sym_2}\rangle, \tag{3.20}$$

and

$$\hat{S}^2 |\psi_{sym_3}\rangle \ = \ 2 |\psi_{sym_3}\rangle. \tag{3.21}$$

It is clear from equations 3.19 through 3.21 that $S(S+1) = 2$ for the symmetric combinations, implying that $S = 1$.

F.

WORK OUT THE RELATIVE TERM ENERGIES FOR TERMS ARISING FROM THE p^2 CONFIGURATION. SHOW THAT

$$\frac{E(^1S) - E(^1D)}{E(^1D) - E(^3P)} = \frac{3}{2}. \tag{3.22}$$

To find the relative term energies, we need to consider only the Coulomb repulsion term $\frac{e^2}{r_{12}}$ and calculate its expectation value for each ^{2S+1}L term. Because space is isotropic and $\frac{e^2}{r_{12}}$ removes only the L degeneracy, we need to calculate the energy of only one M for each L. In addition, because $\frac{e^2}{r_{12}}$ operates only on spatial coordinates, we do not need to consider the spin functions. The matrix elements to be calculated are of the form

$$\langle L \ M | \frac{e^2}{r_{12}} | L \ M \rangle.$$

For simplicity we choose M=L.

The state functions involved in the calculation in the coupled representation are

$$^1S \Rightarrow L = 0, \ M = 0 \Rightarrow |0,0\rangle$$

$$^1D \Rightarrow L = 2, \ M = 2 \Rightarrow |2,2\rangle$$

$$^3P \Rightarrow L = 1, \ M = 1 \Rightarrow |1,1\rangle.$$

In the uncoupled representation, the Coulomb repulsion term becomes

$$\langle n_a, l_a, m_{l_a}(1); n_b, l_b, m_{l_b}(2) | \frac{e^2}{r_{12}} | n_c, l_c, m_{l_c}(1); n_d, l_d, m_{l_d}(2) \rangle$$

$$= \ \sum_k R^k(n_a l_a, n_b l_b, n_c l_c, n_d l_d) \ \frac{4\pi}{2k+1}$$

$$\times \sum_{q=-k}^{k} (-1)^q \langle Y_{l_a m_a} | Y_{k,-q} | Y_{l_c m_c} \rangle \langle Y_{l_b m_b} | Y_{kq} | Y_{l_d m_d} \rangle. \tag{3.23}$$

Because we are dealing with two identical electrons (p^2 configuration), all the n and l values are equal (i.e., $l_a = l_b = l_c = l_d = 1$). We can write

$$R^k(n1, n1, n1, n1) = F^k(n1, n1) = F^k, \tag{3.24}$$

which can be taken out of the brackets, leaving only the spherical harmonic part to be calculated. Before going on, we want to know for which values of k the integrals in equation 3.23 do not vanish. The integral of the product of three spherical harmonics is given by equation Z-3.115,

$$\langle Y_{l_3 m_3} | Y_{l_2 m_2} | Y_{l_1 m_1} \rangle = \sqrt{\frac{(2l_1 + 1)(2l_2 + 1)}{4\pi(2l_3 + 1)}} \langle l_1 0, l_2 0 | l_3 0 \rangle \langle l_1 m_1, l_2 m_2 | l_3 m_3 \rangle \tag{3.25}$$

Comparing equations 3.23 and 3.25, we obtain

$$\langle n, 1, m_{l_a}(1); n, 1, m_{l_b}(2) | \frac{e^2}{r_{12}} | n, 1, m_{l_c}(1); n, 1, m_{l_d}(2) \rangle$$

$$= \sum_k F^k \frac{4\pi}{2k+1} \frac{(2 \times 1 + 1)(2k+1)}{4\pi(2 \times 1 + 1)} (\langle 10, k0 | 10 \rangle)^2$$

$$\times \sum_{q=-k}^{k} (-1)^q \langle 1m_{l_a}, k - q | 1m_{l_c} \rangle \langle 1m_{l_b}, kq | 1m_{l_d} \rangle. \tag{3.26}$$

Because $|1 - k| \leq 1 \leq 1 + k$, it follows that $k = 0, 1$, or 2. The Clebsch-Gordan coefficients that contribute to the sum in equation 3.26 are:

$$k = 0 : \langle 1000 | 10 \rangle = 1,$$

$$k = 1 : \langle 1010 | 10 \rangle = 0,$$

and

$$k = 2 : \langle 1020 | 10 \rangle = \frac{-2}{\sqrt{10}}.$$

The integrals that contribute to the sum are those for $k = 2$ $(q = -2, -1, 0, 1, 2)$ and $k = 0$ $(q = 0)$ only. For the latter, only those with $m_{l_c} = m_{l_a}$ and $m_{l_d} = m_{l_b}$ have nonzero Clebsch-Gordan coefficients.

Going back to the set of states that we want to evaluate, we need to transform the state functions to the uncoupled representation:

$$|0,0\rangle = \langle n, 1, 0; n, 1, 0 | 0, 0 \rangle \ |n, 1, 0; n, 1, 0\rangle + \langle n, 1, 1; n, 1, -1 | 0, 0 \rangle \ |n, 1, 1; n, 1, -1\rangle$$

$$+ \langle n, 1, -1; n, 1, 1 | 0, 0 \rangle \ |n, 1, -1; n, 1, 1\rangle$$

$$= \frac{1}{\sqrt{3}} \left(-|n, 1, 0; n, 1, 0\rangle + |n, 1, 1; n, 1, -1\rangle + |n, 1, -1; n, 1, 1\rangle \right), \tag{3.27}$$

$$|1,1\rangle \; = \; \langle n,1,0;n,1,1\,|\,1,1\rangle \;\; |n,1,0;n,1,1\rangle + \langle n,1,1;n,1,0\,|\,1,1\rangle \;\; |n,1,1;n,1,0\rangle$$

$$= \; \frac{1}{\sqrt{2}}\left(-\,|n,1,0;n,1,1\rangle + |n,1,1;n,1,0\rangle\right), \tag{3.28}$$

and

$$|2,2\rangle \; = \; \langle n,1,1;n,1,1\,|\,2,2\rangle \;\; |n,1,1;n,1,1\rangle$$

$$= \; |n,1,1;n,1,1\rangle. \tag{3.29}$$

We next evaluate the Coulomb matrix elements for each of the above state functions. For ^1S we have:

$$\langle 0,0|\,\frac{e^2}{r_{12}}\,|0,0\rangle \; = \; \frac{1}{3}\,(\; \langle n,1,0;n,1,0|\,\frac{e^2}{r_{12}}\,|n,1,0;n,1,0\rangle - \langle n,1,1;n,1,-1|\,\frac{e^2}{r_{12}}\,|n,1,0;n,1,0\rangle$$

$$- \langle n,1,-1;n,1,1|\,\frac{e^2}{r_{12}}\,|n,1,0;n,1,0\rangle - \langle n,1,0;n,1,0|\,\frac{e^2}{r_{12}}\,|n,1,1;n,1,-1\rangle$$

$$+ \langle n,1,1;n,1,-1|\,\frac{e^2}{r_{12}}\,|n,1,1;n,1,-1\rangle + \langle n,1,-1;n,1,1|\,\frac{e^2}{r_{12}}\,|n,1,1;n,1,-1\rangle$$

$$- \langle n,1,0;n,1,0|\,\frac{e^2}{r_{12}}\,|n,1,-1;n,1,1\rangle + \langle n,1,1;n,1,-1|\,\frac{e^2}{r_{12}}\,|n,1,-1;n,1,1\rangle$$

$$+ \langle n,1,-1;n,1,1|\,\frac{e^2}{r_{12}}\,|n,1,-1;n,1,1\rangle \;). \tag{3.30}$$

Evaluating each of these terms using equation 3.26, we obtain

$$\langle n,1,0;n,1,0|\,\frac{e^2}{r_{12}}\,|n,1,0;n,1,0\rangle \; = \; F^2 \;\langle 1020|10\rangle^2 \sum_{q=-2}^{2}(-1)^q\,\langle 102-q|10\rangle\,\langle 102q|10\rangle$$

$$+F^0 \;\langle 1000|10\rangle^2 \,\langle 1000|10\rangle\,\langle 1000|10\rangle$$

$$= \; F^2\left(\frac{-2}{\sqrt{10}}\right)^2 [\; 2\,\langle 1022|10\rangle\,\langle 102-2|10\rangle$$

$$-2\,\langle 1021|10\rangle\,\langle 102-1|10\rangle + \langle 1020|10\rangle\,\langle 1020|10\rangle\,] + F^0$$

$$= F^2 \left(\frac{-2}{\sqrt{10}}\right)^2 \left(\frac{-2}{\sqrt{10}}\right)^2 + F^0$$

$$= \frac{4}{25}F^2 + F^0, \tag{3.31}$$

$$\langle n,1,1;n,1,-1|\frac{e^2}{r_{12}}|n,1,0;n,1,0\rangle = \frac{2}{5}F^2 \sum_{q=-2}^{2}(-1)^q \langle 112-q|10\rangle \langle 1-12q|10\rangle$$

$$= \frac{2}{5}F^2(-1)\langle 112-1|10\rangle \langle 1-121|10\rangle$$

$$= -\frac{3}{25}F^2, \tag{3.32}$$

$$\langle n,1,-1;n,1,1|\frac{e^2}{r_{12}}|n,1,0;n,1,0\rangle = \frac{2}{5}F^2 \sum_{q=-2}^{2}(-1)^q \langle 1-12-q|10\rangle \langle 112q|10\rangle$$

$$= \frac{2}{5}F^2(-1)\langle 1-121|10\rangle \langle 112-1|10\rangle$$

$$= -\frac{3}{25}F^2, \tag{3.33}$$

$$\langle n,1,0;n,1,0|\frac{e^2}{r_{12}}|n,1,1;n,1,-1\rangle = \langle n,1,1;n,1,-1|\frac{e^2}{r_{12}}|n,1,0;n,1,0\rangle^*$$

$$= -\frac{3}{25}F^2, \tag{3.34}$$

where we have used the fact that F^k is real.

$$\langle n,1,1;n,1,-1|\frac{e^2}{r_{12}}|n,1,1;n,1,-1\rangle = \frac{2}{5}F^2 \sum_{q=-2}^{2}(-1)^q \langle 112-q|11\rangle \langle 1-12q|1-1\rangle$$

$$+F^0 \langle 1100|11\rangle \langle 1-100|1-1\rangle$$

$$= \frac{2}{5}F^2 \langle 1120|11\rangle \langle 1-120|1-1\rangle + F^0$$

$$= \frac{1}{25}F^2 + F^0, \tag{3.35}$$

$$\langle n,1,-1;n,1,1| \frac{e^2}{r_{12}} |n,1,1;n,1,-1\rangle = \frac{2}{5}F^2 \sum_{q=-2}^{2} (-1)^q \langle 1-12-q|11\rangle \langle 112q|1-1\rangle$$

$$= \frac{2}{5}F^2 \langle 1-122|11\rangle \langle 112-2|1-1\rangle$$

$$= \frac{6}{25}F^2, \tag{3.36}$$

$$\langle n,1,0;n,1,0| \frac{e^2}{r_{12}} |n,1,-1;n,1,1\rangle = \langle n,1,-1;n,1,1| \frac{e^2}{r_{12}} |n,1,0;n,1,0\rangle^*$$

$$= -\frac{3}{25}F^2, \tag{3.37}$$

$$\langle n,1,1;n,1,-1| \frac{e^2}{r_{12}} |n,1,-1;n,1,1\rangle = \langle n,1,-1;n,1,1| \frac{e^2}{r_{12}} |n,1,1;n,1,-1\rangle^*$$

$$= \frac{6}{25}F^2, \tag{3.38}$$

and

$$\langle n,1,-1;n,1,1| \frac{e^2}{r_{12}} |n,1,-1;n,1,1\rangle = \frac{2}{5}F^2 \sum_{q=-2}^{2} (-1)^q \langle 1-12-q|1-1\rangle \langle 112q|11\rangle$$

$$+F^0 \langle 1-100|1-1\rangle \langle 1100|11\rangle$$

$$= \frac{2}{5}F^2 \langle 1-120|1-1\rangle \langle 1120|11\rangle + F^0$$

$$= \frac{1}{25}F^2 + F^0. \tag{3.39}$$

Putting all these results together into equation 3.30, we get

$$\langle {}^1S| r_{12}^{-1} |{}^1S\rangle = \frac{1}{3}\left(\frac{4}{25}F^2 + F^0 + \frac{3}{25}F^2 + \frac{3}{25}F^2 + \frac{3}{25}F^2 + \frac{1}{25}F^2 + F^0 + \frac{6}{25}F^2 \right.$$

$$+\frac{3}{25}F^2 + \frac{6}{25}F^2 + \frac{1}{25}F^2 + F^0 \Bigg)$$

$$= \frac{2}{5}F^2 + F^0. \tag{3.40}$$

For ^1D, we obtain

$$\langle 2,2|\frac{e^2}{r_{12}}|2,2\rangle = \langle n,1,1;n,1,1|\frac{e^2}{r_{12}}|n,1,1;n,1,1\rangle$$

$$= \frac{2}{5}F^2 \sum_{q=-2}^{2} (-1)^q \langle 112-q|11\rangle \langle 112q|11\rangle$$

$$+F^0 \langle 1100|11\rangle \langle 1100|11\rangle$$

$$= \frac{2}{5}F^2 \langle 1120|11\rangle \langle 1120|11\rangle + F^0$$

$$= \frac{1}{25}F^2 + F^0. \tag{3.41}$$

For ^3P, we obtain

$$\langle 1,1|\frac{e^2}{r_{12}}|1,1\rangle = \frac{1}{2}\Bigg(\langle n,1,0;n,1,1|\frac{e^2}{r_{12}}|n,1,0;n,1,1\rangle - \langle n,1,0;n,1,1|\frac{e^2}{r_{12}}|n,1,1;n,1,0\rangle$$

$$-\langle n,1,1;n,1,0|\frac{e^2}{r_{12}}|n,1,0;n,1,1\rangle + \langle n,1,1;n,1,0|\frac{e^2}{r_{12}}|n,1,1;n,1,0\rangle\Bigg). \tag{3.42}$$

Calculating each of the matrix elements using 3.23, we obtain

$$\langle n,1,0;n,1,1|\frac{e^2}{r_{12}}|n,1,0;n,1,1\rangle = \frac{2}{5}F^2 \sum_{q=-2}^{2} (-1)^q \langle 102-q|10\rangle \langle 112q|11\rangle$$

$$+F^0 \langle 1000|10\rangle \langle 1100|11\rangle$$

$$= \frac{2}{5}F^2 \langle 1020|10\rangle \langle 1120|11\rangle + F^0$$

$$= -\frac{2}{25}F^2 + F^0 \tag{3.43}$$

$$\langle n,1,0;n,1,1| \frac{e^2}{r_{12}} |n,1,1;n,1,0\rangle = \frac{2}{5}F^2 \sum_{q=-2}^{2} (-1)^q \langle 102-q|11\rangle \langle 112q|10\rangle$$

$$= \frac{2}{5}F^2(-1) \langle 1021|11\rangle \langle 112-1|10\rangle$$

$$= \frac{3}{25}F^2, \tag{3.44}$$

$$\langle n,1,1;n,1,0| r\frac{e^2}{r_{12}} |n,1,0;n,1,1\rangle = \langle n,1,0;n,1,1| \frac{e^2}{r_{12}} |n,1,1;n,1,0\rangle^*$$

$$= \frac{3}{25}F^2, \tag{3.45}$$

and

$$\langle n,1,1;n,1,0| \frac{e^2}{r_{12}} |n,1,1;n,1,0\rangle = \frac{2}{5}F^2 \sum_{q=-2}^{2} (-1)^q \langle 112-q|11\rangle \langle 102q|10\rangle$$

$$+F^0 \, \langle 1100|11\rangle \langle 1000|10\rangle$$

$$= \frac{2}{5}F^2 \, \langle 1120|11\rangle \langle 1020|10\rangle + F^0$$

$$= -\frac{2}{25}F^2 + F^0. \tag{3.46}$$

Finally, combining equations 3.43 through 3.46, we obtain

$$\langle {}^3P| \frac{e^2}{r_{12}} |{}^3P\rangle = \frac{1}{2} \left(-\frac{2}{25}F^2 + F^0 - \frac{3}{25}F^2 - \frac{3}{25}F^2 - \frac{2}{25}F^2 + F^0 \right)$$

$$= -\frac{1}{5}F^2 + F^0. \tag{3.47}$$

To complete the calculation, we substitute equations 3.40, 3.41, and 3.47 into equation 3.22 and obtain

$$\frac{E({}^1S) - E({}^1D)}{E({}^1D) - E({}^3P)} = \frac{F^0 + \frac{2}{5}F^2 - F^0 - \frac{1}{25}F^2}{F^0 + \frac{1}{25}F^2 - F^0 + \frac{1}{5}F^2}$$

$$= \frac{3}{2}. \tag{3.48}$$

APPLICATION 5
ANGULAR DISTRIBUTION OF RIGID ROTOR AXES FOLLOWING ABSORPTION OF PLANE-POLARIZED LIGHT

THE PROBABILITY *amplitude* OF FINDING THE RIGID ROTOR IN THE STATE $|JM\rangle$ FOLLOWING THE DIPOLE ABSORPTION OF PLANE-POLARIZED RADIATION ($j_{ph} = 1, m_{ph} = 0$) IS PROPORTIONAL TO $\langle J''M'', 10| JM\rangle$, WHERE $|J''M''\rangle$ IS THE INITIAL STATE OF THE ROTOR. THE M'' STATES ARE ASSUMED TO BE EQUALLY POPULATED WITH NO PHASE RELATIONS AMONG THEM. THUS, THE TOTAL PROBABILITY OF FINDING THE ROTOR AXIS POINTING INTO THE SOLID ANGLE ELEMENT $d\Omega$ IS OBTAINED BY SIMPLY SUMMING OVER ALL INITIAL M'' STATES, GIVING

$$P_J(\theta) = \sum_{M''} \langle J''M'', 10| JM\rangle^2 |Y_{JM}(\theta, \phi)|^2 . \tag{3.49}$$

WE WANT TO EVALUATE EQUATION 3.49.

A.

USE THE CLEBSCH-GORDAN SERIES OR ITS INVERSE TO SHOW THAT

$$\langle j_1 m_1, j_2 m_2 | j_3 m_3\rangle D^{j_3}_{k_3 m_3} = \sum_{k_1} \langle j_1 k_1, j_2 k_3 - k_1 | j_3 k_3\rangle D^{j_1}_{k_1 m_1} D^{j_2}_{k_3 - k_1 m_2}. \tag{3.50}$$

We start with the Clebsch-Gordan transformation,

$$|j_1 m_1, j_2 m_2\rangle = \sum_{j_3} \langle j_3 m_3 | j_1 m_1, j_2 m_2\rangle |j_3 m_3\rangle . \tag{3.51}$$

Applying a rotation of the form

$$\mathbf{R}(\phi, \theta, \chi) |jm\rangle = \sum_k D^j_{km}(\phi, \theta, \chi) |jk\rangle \tag{3.52}$$

to both sides of equation 3.51 gives the relation

$$\sum_{k_1} \sum_{k_2} D^{j_1}_{k_1 m_1} D^{j_2}_{k_2 m_2} |j_1 k_1, j_2 k_2\rangle = \sum_{k_3} \sum_{j_3} \langle j_3 m_3 | j_1 m_1, j_2 m_2\rangle D^{j_3}_{k_3 m_3} |j_3 k_3\rangle . \tag{3.53}$$

Multiplying both sides of this equation by a specific $\langle j_3 k_3|$ state, where $k_3 = k_1 + k_2$,

$$\sum_{k_1} D^{j_1}_{k_1 m_1} D^{j_2}_{k_3 - k_1 m_2} \langle j_3 k_3 | j_1 k_1, j_2 k_3 - k_1\rangle = \langle j_3 m_3 | j_1 m_1, j_2 m_2\rangle D^{j_3}_{k_3 m_3}, \tag{3.54}$$

we obtain the equality that we are looking for.

B.

MAKE THE IDENTIFICATION

$$
\begin{array}{lll}
j_1 = J'' & m_1 = M'' & k_1 = K'' \\
j_2 = 1 & m_2 = 0 & k_2 = K - K'' \\
j_3 = J & m_3 = M & k_3 = K.
\end{array}
$$

THEN EQUATION 3.54 BECOMES

$$\sum_{K''} \langle J''K'', 1K - K'' | JK \rangle D_{K''M''}^{J''} D_{K-K''0}^{1} = \langle J''M'', 10 | JM \rangle D_{KM}^{J}.$$ (3.55)

SQUARE BOTH SIDES AND SUM OVER M'' TO PROVE THAT

$$\sum_{K''} \langle JK | J''K'', 1K - K'' \rangle^2 \left| D_{K-K''0}^{1} \right|^2 = \sum_{M''} \left| D_{KM}^{J} \right|^2 \left| \langle J''M'', 10 | JM \rangle \right|^2$$ (3.56)

WHERE THE SUM OVER K'' IS FROM K-1 TO K+1.

Squaring both sides of equation 3.55 and summing over M'', we obtain

$$\sum_{M''} \left| \langle J''M'', 10 | JM \rangle \right|^2 \left| D_{KM}^{J} \right|^2 = \sum_{M''} \sum_{K''} \sum_{k''} \left[\langle J''K'', 1K - K'' | JK \rangle \langle J''k'', 1K - k'' | JK \rangle \right.$$

$$\left. \times D_{K''M''}^{J''} D_{K-K''0}^{1} D_{k''M}^{J} D_{K-k''0}^{1} \right].$$ (3.57)

The summation on the right side of equation 3.57 consists of terms for which $K'' = k''$ and terms for which $K'' \neq k''$. For those terms with $K'' = k''$, we obtain

$$\sum_{M''} \langle JK | J''K'', 1K - K'' \rangle^2 \left| D_{K''M''}^{J''} \right|^2 \left| D_{K-K''0}^{1} \right|^2 = \langle JK | J''K'', 1K - K'' \rangle^2 \left| D_{K-K''0}^{1} \right|^2$$ (3.58)

because

$$\sum_{M''} \left| D_{K''M''}^{J''} \right|^2 = 1.$$ (3.59)

For those terms with $K'' \neq k''$, we get

$$\sum_{M''} \langle JK | J''K'', 1K - K'' \rangle \langle JK | J''k'', 1K - k'' \rangle D_{K''M''}^{J''} D_{K-K''0}^{1} D_{k''M''}^{J''} D_{K-k''0}^{1}$$

$$= \langle JK | J''K'', 1K - K'' \rangle \langle JK | J''k'', 1K - k'' \rangle D_{K-K''0}^{1} D_{K-k''0}^{1} \sum_{M''} D_{K''M''}^{J''} D_{k''M''}^{J''}$$

$$= 0,$$ (3.60)

where we used

$$\sum_{M''} D_{K''M''}^{J''} D_{k''M''}^{J''} = \delta_{K''k''}.$$ (3.61)

Substituting equations 3.58 and 3.60 into equation 3.57, we obtain the desired result. The sum over K'' is from $K - 1$ to $K + 1$. Thus, equation 3.56 has the explicit value

$$\sum_{M''} \left| D_{KM}^{J} \right|^2 \left| \langle J''M'', 10 | JM \rangle \right|^2 = \langle J''K - 1, 11 | JK \rangle^2 \left| D_{10}^{1} \right|^2$$

$$+ \langle J''K, 10 | JK \rangle^2 \left| D_{00}^{1} \right|^2 + \langle J''K + 1, 1 - 1 | JK \rangle^2 \left| D_{-10}^{1} \right|^2$$ (3.62)

C.

USE THE FACT THAT $\left|D_{KM}^{J}\right|^{2}$ FOR $K = 0$ IS EQUAL TO $[4\pi/(2J+1)]\left|Y_{JM}\right|^{2}$ TO EXPRESS $P_{J}(\theta)$ IN THE FORM

$$P_{J}(\theta) = a\sin^{2}\theta + b\cos^{2}\theta. \tag{3.63}$$

FIND THE COEFFICIENTS a AND b AS AN EXPLICIT FUNCTION OF J'' FOR THE TWO TYPES OF TRANSITIONS, A P-BRANCH IN WHICH $\Delta J = J - J'' = -1$ AND AN R-BRANCH IN WHICH $\Delta J = J - J'' = +1$.

We know from equations Z-3.93 and Z-3.97 that

$$\left|D_{0M}^{J}\right|^{2} = \frac{4\pi}{2J+1}\left|Y_{JM}\right|^{2} \tag{3.64}$$

and

$$\left|D_{-10}^{1}\right|^{2} = \left|D_{10}^{1}\right|^{2}$$

$$= \frac{4\pi}{3}\left|Y_{11}\right|^{2}. \tag{3.65}$$

Evaluating equation 3.62 for $K = 0$, we obtain

$$\sum_{M''}\left|D_{0M}^{J}\right|^{2}\left|\langle J''M'', 10\,|JM\rangle\right|^{2}$$

$$= \left|D_{10}^{1}\right|^{2}\left[\langle J''-1, 11\,|J0\rangle^{2} + \langle J''1, 1-1\,|J0\rangle^{2}\right] + \langle J''0, 10\,|J0\rangle^{2}\left|D_{00}^{1}\right|^{2}. \tag{3.66}$$

Inserting equations 3.64 and 3.65 into equation 3.66, we get

$$\frac{4\pi}{2J+1}\sum_{M''}\left|\langle J''M'', 10\,|JM\rangle\right|^{2}\left|Y_{JM}\right|^{2} = \frac{4\pi}{3}\left[\left|Y_{11}\right|^{2}\left(\langle J''-1, 11\,|J0\rangle^{2} + \langle J''1, 1-1\,|J0\rangle^{2}\right)\right.$$

$$\left. + \left|Y_{10}\right|^{2}\langle J''0, 10|J0\rangle^{2}\right]$$

$$= \frac{\sin^{2}\theta}{2}\left[\langle J''-1, 11\,|J0\rangle^{2} + \langle J''1, 1-1\,|J0\rangle^{2}\right]$$

$$+ \cos^{2}\theta\,\langle J''0, 10\,|J0\rangle^{2}. \tag{3.67}$$

The left side of equation 3.67 is proportional to $P_{J}(\theta)$ (defined in equation 3.49). Because

$$\langle J''-1, 11\,|J0\rangle^{2} = \langle J''1, 1-1\,|J0\rangle^{2}, \tag{3.68}$$

we get

$$P_J(\theta) = \frac{2J+1}{4\pi}\left(\langle J''1,1-1\,|J0\rangle^2\sin^2\theta + \langle J''0,10|\,J0\rangle^2\cos^2\theta\right), \tag{3.69}$$

or more compactly

$$P_J(\theta) = a\sin^2\theta + b\cos^2\theta, \tag{3.70}$$

where

$$a = \frac{(2J+1)\,\langle J''1,1-1\,|J0\rangle^2}{4\pi}, \tag{3.71}$$

and

$$b = \frac{(2J+1)\,\langle J''0,10\,|J0\rangle^2}{4\pi}. \tag{3.72}$$

For a P-branch transition, where $\Delta J = J - J'' = -1$, we obtain for a and b

$$a = \frac{[2(J''-1)+1]\,|\langle J''1,1-1\,|J''-10\rangle|^2}{4\pi}$$

$$= \frac{J''(J''+1)(2J''-1)}{4\pi(2J'')(2J''+1)}$$

$$= \frac{(J''+1)(2J''-1)}{8\pi(2J''+1)} \tag{3.73}$$

and

$$b = \frac{2J+1\,\langle J''0,10\,|J''-10\rangle^2}{4\pi}$$

$$= \frac{J''(2J''-1)}{4\pi(2J''+1)}. \tag{3.74}$$

For an R-branch transition, where $\Delta J = +1$, we obtain

$$a = \frac{J''(J''+1)[2(J''+1)+1]}{4\pi(2J''+1)(2J''+2)}$$

$$= \frac{J''(2J''+3)}{8\pi(2J''+1)} \tag{3.75}$$

and

$$b = \frac{(J''+1)^2(2J''+3)}{4\pi(2J''+1)(J''+1)}$$

$$= \frac{(J''+1)(2J''+3)}{4\pi(2J''+1)}. \tag{3.76}$$

D.

$P_J(\theta)$ CAN BE RECAST INTO THE FAMILIAR FORM

$$P_J(\theta) = \frac{1 + \mathcal{A}_0(J)P_2(\cos\theta)}{4\pi}, \tag{3.77}$$

WHERE

$$\mathcal{A}_0 = \frac{2(b-a)}{2a+b} \tag{3.78}$$

AND

$$P_2(\cos\theta) = \frac{3\cos^2\theta - 1}{2}. \tag{3.79}$$

SHOW THAT $\mathcal{A}_0 \to \frac{1}{2}$ FOR EITHER BRANCH AS $J \to \infty$.

For a P-branch $\mathcal{A}_0$ becomes

$$\mathcal{A}_0 = \left[\frac{2J''(2J''-1)}{4\pi(2J''+1)} - \frac{2(J''+1)(2J''-1)}{8\pi(2J''+1)}\right]\left[\frac{2(J''+1)(2J''-1)}{8\pi(2J''+1)} + \frac{J''(2J''-1)}{4\pi(2J''+1)}\right]^{-1}$$

$$= \frac{J''-1}{2J''+1}. \tag{3.80}$$

Thus, we obtain

$$\lim_{J''\to\infty}\mathcal{A}_0 = \lim_{J''\to\infty}\frac{J''-1}{2J''+1}$$

$$= \frac{1}{2}. \tag{3.81}$$

For an R-branch, we get

$$\mathcal{A}_0 = \left[\frac{2(J''+1)(2J''+3)}{4\pi(2J''+1)} - \frac{2J''(J''+3)}{8\pi(2J''+1)}\right]\left[\frac{2J''(2J''+3)}{8\pi(2J''+1)} + \frac{(J''+1)(2J''+3)}{4\pi(2J''+1)}\right]^{-1}$$

$$= \frac{J''+2}{2J''+1}. \tag{3.82}$$

Again we obtain

$$\lim_{J''\to\infty}\mathcal{A}_0 = \lim_{J''\to\infty}\frac{J''+2}{2J''+1}$$

$$= \frac{1}{2}. \tag{3.83}$$

$\mathcal{A}_0$ is called the *alignment parameter*. It ranges in value from $+2$ for a pure $\cos^2\theta$ distribution to -1 for a pure $\sin^2\theta$ distribution. For $\mathcal{A}_0 = 0$, the distribution is isotropic. This problem illustrates the use of plane-polarized light to prepare aligned molecules that may then be used as projectiles or targets in scattering experiments. It is a special case of the production of anisotropically distributed targets by beam excitation.

ADDENDUM

These results can be generalized[1] to transitions of a symmetric top by writing equation 3.49 in the more general form

$$P_{JK}(\theta) \quad \propto \quad \frac{2J+1}{8\pi} \sum_M \sum_{M''} P(JKM, J''K''M'') \left| D_{MK}^{J*}(\phi, \theta, \chi) \right|^2, \tag{3.84}$$

where $P(JKM, J''K''M'')$ is the probability for a transition from $|J''K''M''\rangle$ to $|JKM\rangle$, and

$$|JKM\rangle = \sqrt{\frac{2J+1}{8\pi}} D_{MK}^J(\phi, \theta, \chi)^2$$

is the wave function of the final state. The probability for a one-photon transition in a weak field is given by

$$P(J, K, M; J'', K'', M'') \propto \left| \langle JKM | \vec{\mu} \cdot \vec{E} | J''K''M'' \rangle \right|^2, \tag{3.85}$$

where $\vec{\mu}$ is the transition dipole operator and $\vec{E}$ is the electric field.

We set the polarization direction of the photon to coincide with the Z axis in the laboratory frame. The component of $\vec{\mu}$ along this direction is $\mu(1,0)$. We can relate $\mu(1,0)$ to its value in the molecular frame by

$$\mu(1,0) = \sum_{q=-1}^{1} D_{0q}^{1*}(R)\mu(1,q), \tag{3.86}$$

where $q = \pm 1$ corresponds to a perpendicular transition and $q = 0$ corresponds to a parallel transition. Substituting equations 3.85 and 3.86 into equation 3.84 gives

$$P_{JK}(\theta) \quad \propto \quad \sum_M \sum_{M''} \left| \left\langle JKM \left| \sum_q D_{0q}^{1*}(R)\mu(1,q) \right| J''K''M'' \right\rangle \right|^2 \times \left| D_{KM}^J(\phi, \theta, \chi) \right|^2. \tag{3.87}$$

The first factor in the sum can be evaluated using equation Z-3.118:

$$\left| \sum_q \langle JKM | D_{0q}^{1*}(R)\mu(1,q) | J''K''M'' \rangle \right|^2$$

$$= \frac{1}{8\pi^2}(2J+1)(2J''+1) \left| \sum_q \mu(1,q) \int_{d\Omega} D_{MK}^J(R) D_{0q}^{1*}(R) D_{M''K''}^{J''*}(R) \right|^2$$

---

[1] R. Liyanage, and R. J. Gordon, *J. Chem. Phys.* **107**, 7209 (1997).

$$= \frac{1}{8\pi^2}(2J+1)(2J''+1)\left|\sum_q (-1)^{-q+M''-K''}\mu(1,q)\right.$$

$$\left.\times \int_{d\Omega} D^J_{MK}(R)D^1_{0-q}(R)D^{J''}_{-M''-K''}(R)\right|^2$$

$$= (2J+1)(2J''+1)\left|\sum_q \mu(1,q)\begin{pmatrix} J'' & 1 & J \\ -M'' & 0 & M \end{pmatrix}\begin{pmatrix} J'' & 1 & J \\ -K'' & -q & K \end{pmatrix}\right|^2. \tag{3.88}$$

For a particular choice of K and K'', only the term with $q = K - K''$ survives, and

$$P(J,K,M;J'',K'',M'') \propto \langle J''M'',10|JM\rangle^2 \langle J''K'',1q|JK\rangle^2. \tag{3.89}$$

The angular distribution is therefore given by

$$P_{JK}(\theta) \propto \langle J''K'',1q|JK\rangle^2 \sum_M \sum_{M''} \langle J''M'',10|JM\rangle^2 |D^J_{KM}(\phi,\theta,\chi)|^2. \tag{3.90}$$

We recognize the quantity $\langle J''K'',1q|JK\rangle^2$ to be proportional to the Hönl-London factor for a symmetric top transition (equation Z-6.117). The generalization of equation 3.84 to multiphoton transitions and to an asymmetric top is straightforward.

APPLICATION 6
PHOTOFRAGMENT ANGULAR DISTRIBUTION
(CLASSICAL TREATMENT)

A.

SHOW THAT THE FRAGMENT ANGULAR DISTRIBUTION HAS THE FORM

$$I(\theta_s, \phi_s) = \frac{\sigma}{4\pi} \left[1 + \beta P_2(\cos \theta_s)\right], \tag{3.91}$$

WHERE

$$\sigma = (4\pi)^{\frac{1}{2}} b_{00}$$

$$= \int_0^{2\pi} \mathrm{d}\phi_m \int_0^{\pi} \sin \theta_m \mathrm{d}\theta_m f(\theta_m, \phi_m) \tag{3.92}$$

IS THE TOTAL CROSS SECTION, AND

$$\beta = \frac{2b_{20}}{\sqrt{5} b_{00}}$$

$$= \frac{2 \int_0^{2\pi} \mathrm{d}\phi_m \int_0^{\pi} \sin \theta_m \mathrm{d}\theta_m P_2(\cos \theta_m) f(\theta_m, \phi_m)}{\int_0^{2\pi} \mathrm{d}\phi_m \int_0^{\pi} \sin \theta_m \mathrm{d}\theta_m f(\theta_m, \phi_m)}$$

$$= 2 \langle P_2(\cos \theta_m) \rangle, \tag{3.93}$$

IS THE ASYMMETRY PARAMETER. HERE $\langle \cdots \rangle$ DENOTES AN AVERAGE OVER THE RECOIL DISTRIBUTION $f(\theta_m, \phi_m)$.

The fragment angular distribution in the laboratory frame is given by

$$I(\theta_s, \phi_s) = \int_0^{2\pi} \mathrm{d}\phi \int_0^{\pi} \sin \theta \mathrm{d}\theta \int_0^{2\pi} \mathrm{d}\chi P_{\mathrm{diss}}(\phi, \theta, \chi) f(\theta_m, \phi_m), \tag{3.94}$$

where P_{diss} designates the probability of making a dissociative transition,

$$P_{\mathrm{diss}} = \frac{1}{8\pi^2} \left[D_{00}^0(\phi, \theta, \chi) + 2D_{00}^2(\phi, \theta, \chi) \right], \tag{3.95}$$

$f(\theta_m, \phi_m)$ is the fragment recoil distribution in the molecular frame, and ϕ, θ, and χ, are the Euler angles describing the frame transformation from molecular to laboratory coordinates. Whatever its form, $f(\theta_m, \phi_m)$ can always be expanded in terms of spherical harmonics that form a complete set,

$$f(\theta_m, \phi_m) = \sum_{k,q} b_{kq} Y_{kq}(\theta_m, \phi_m), \tag{3.96}$$

where the expansion coefficient b_{kq} is given by

$$b_{kq} = \int_0^{2\pi} d\phi_m \int_0^\pi \sin\theta_m d\theta_m \int_0^{2\pi} d\chi Y_{kq}^*(\theta_m,\phi_m) f(\theta_m,\phi_m). \tag{3.97}$$

We start by replacing the final recoil distribution in the *molecular frame* by its equivalent function in the *laboratory frame*. Inserting equations 3.95 and 3.96 into equation 3.94, we obtain

$$I(\theta_s,\phi_s) = \int_0^{2\pi}\int_0^\pi\int_0^{2\pi} \frac{1}{8\pi^2} \left[D_{00}^0(\phi,\theta,\chi) + 2D_{00}^2(\phi,\theta,\chi) \right] \sum_{k,q} b_{kq} Y_{kq}(\theta_m,\phi_m) d\phi \sin\theta d\theta d\chi.$$

$$\tag{3.98}$$

The spherical harmonic $Y_{kq}(\theta_m,\phi_m)$ can be expressed in the laboratory frame using the frame rotation from molecular to laboratory coordinates,

$$Y_{kq}(\theta_m,\phi_m) = \sum_M D_{Mq}^k(\phi,\theta,\chi) Y_{kM}(\theta_s,\phi_s), \tag{3.99}$$

which can be inserted into equation 3.98 to obtain

$$I(\theta_s,\phi_s) = \sum_{kqM} \frac{1}{8\pi^2} b_{kq} Y_{kM}(\theta_s,\phi_s) \int [D_{00}^0(R)D_{Mq}^k(R) + 2D_{00}^2(R)D_{Mq}^k(R)]d\Omega. \tag{3.100}$$

Using the properties of the rotation matrices,

$$D_{00}^L = D_{00}^{L*} \tag{3.101}$$

and

$$\int d\Omega D_{k_1 m_1}^{j_1 *}(R) D_{k_2 m_2}^{j_2}(R) = \frac{8\pi^2}{2j_1+1} \delta_{j_1 j_2}\delta_{k_1 k_2}\delta_{m_1 m_2}, \tag{3.102}$$

we can evaluate the integral in equation 3.100 to obtain

$$I(\theta_s,\phi_s) = b_{00}Y_{00}(\theta_s,\phi_s) + \frac{2}{5}b_{20}Y_{20}(\theta_s,\phi_s)$$

$$= \frac{b_{00}}{\sqrt{4\pi}}[1 + \frac{2b_{20}}{\sqrt{5}b_{00}} P_2(\cos\theta_s)]. \tag{3.103}$$

By defining

$$\sigma = \sqrt{4\pi}b_{00}, \tag{3.104}$$

and

$$\beta = \frac{2b_{20}}{\sqrt{5}b_{00}}, \tag{3.105}$$

we obtain the fragment angular distribution in the form

$$I(\theta_s,\phi_s) = \frac{\sigma}{4\pi}\left[1 + \beta P_2(\cos\theta_s)\right]. \tag{3.106}$$

B.

CONSIDER ONCE MORE THE LIMITING CASE OF DIATOMIC PHOTODISSOCIATION IN THE AXIAL RECOIL APPROXIMATION. SHOW THAT $\beta = 2$, THAT IS, THAT $I(\theta_s, \phi_s)$ IS PROPORTIONAL TO $\cos^2 \theta_s$, FOR A PARALLEL-TYPE TRANSITION (μ PARALLEL TO THE INTERNUCLEAR AXIS), WHILE $\beta = -1$, THAT IS, $I(\theta_s, \phi_s)$ IS PROPORTIONAL TO $\sin^2 \theta_s$, FOR A PERPENDICULAR-TYPE TRANSITION (μ PERPENDICULAR TO THE INTERNUCLEAR AXIS).

In the axial recoil approximation, in which we assume that dissociation occurs much faster than rotation, $f(\theta_m, \phi_m)$ takes the form

$$f(\theta_m, \phi_m) = \frac{\delta(\theta_m - \theta_m^0)}{2\pi \sin \theta_m}. \tag{3.107}$$

When the transition dipole moment μ is parallel to the internuclear axis (i.e., for a parallel-type transition) $\theta_m^0 = 0$, whereas when μ is perpendicular to the axis (for a perpendicular-type transition), $\theta_m^0 = \pi/2$. Substituting equation 3.107 into equation 3.97, we can evaluate b_{kq},

$$\begin{aligned}
b_{00} &= \int_0^{2\pi} \int_0^\pi Y_{00}^* \frac{\delta(\theta_m - \theta_m^0)}{2\pi \sin \theta_m} \sin \theta_m \mathrm{d}\theta_m \mathrm{d}\phi_m \\
\\
&= 2\pi \frac{1}{\sqrt{4\pi}} \frac{1}{2\pi} \int_0^\pi \delta(\theta_m - \theta_m^0) \mathrm{d}\theta_m \\
\\
&= \frac{1}{\sqrt{4\pi}},
\end{aligned} \tag{3.108}$$

and

$$\begin{aligned}
b_{20} &= \int_0^{2\pi} \int_0^\pi Y_{20}^*(\theta_m, \phi_m) \frac{\delta(\theta_m - \theta_m^0)}{2\pi} \mathrm{d}\theta_m \mathrm{d}\phi_m \\
\\
&= \sqrt{\frac{5}{4\pi}} P_2(\cos \theta_m^0) \\
\\
&= \sqrt{\frac{5}{16\pi}} (3\cos^2 \theta_m^0 - 1).
\end{aligned} \tag{3.109}$$

For $\theta_m^0 = 0$ (the parallel-type transition),

$$\begin{aligned}
\beta &= \frac{2\sqrt{\frac{5}{16\pi}}(3\cos^2 0 - 1)}{\sqrt{\frac{5}{4\pi}}} \\
\\
&= 2.
\end{aligned} \tag{3.110}$$

Substituting β into equation 3.106, we obtain

$$I(\theta_s, \phi_s) = \frac{\sigma}{4\pi}[1 + 2P_2(\cos\theta_s)]$$

$$= \frac{3\sigma}{4\pi}\cos^2\theta_s. \tag{3.111}$$

For $\theta_m^0 = \frac{\pi}{2}$ (perpendicular-type transition),

$$\beta = \frac{2\sqrt{\frac{5}{16\pi}}(3\cos^2\frac{\pi}{2} - 1)}{\sqrt{5} \times \frac{1}{\sqrt{4\pi}}}$$

$$= -1. \tag{3.112}$$

Substituting β into equation 3.106, we obtain

$$I(\theta_s, \phi_s) = \frac{\sigma}{4\pi}[1 + (-1)P_2(\cos\theta_s)]$$

$$= \frac{3\sigma}{8\pi}(1 - \cos^2\theta_s)$$

$$= \frac{3\sigma}{8\pi}\sin^2\theta_s. \tag{3.113}$$

APPLICATION 7
INTRODUCTION TO SYMMETRIC TOPS

CONSIDER A SYMMETRIC TOP WITH TOTAL ANGULAR MOMENTUM $\mathbf{J}$ THAT MAKES A PROJECTION M ON THE SPACE-FIXED Z AXIS AND A PROJECTION K ON THE BODY-FIXED z AXIS. LET ψ_{JKM} DENOTE ITS WAVE FUNCTION. BY SYMMETRY, ITS PERMANENT DIPOLE MOMENT (IF IT HAS ONE) MUST LIE ALONG z. THEN THE PROBABILITY OF THE DIPOLE MOMENT POINTING BETWEEN θ AND $\theta + \mathrm{d}\theta$ IS GIVEN BY

$$P_{JKM}(\theta)\sin\theta\mathrm{d}\theta \;\; = \;\; \left[\int_0^{2\pi}\mathrm{d}\phi \int_0^{2\pi}\mathrm{d}\chi \psi^*_{JKM}(\phi,\theta,\chi)\psi_{JKM}(\phi,\theta,\chi) \right] \sin\theta\mathrm{d}\theta, \qquad (3.114)$$

WHERE θ IS THE ANGLE BETWEEN THE z AND THE Z AXES.

A.

SHOW THAT

$$P_{JKM}(\theta) = (-1)^{M-K}\;\frac{2J+1}{2}\sum_{n=0}^{2J}(2n+1) \begin{pmatrix} J & J & n \\ M & -M & 0 \end{pmatrix} \begin{pmatrix} J & J & n \\ K & -K & 0 \end{pmatrix} P_n(\cos\theta), \qquad (3.115)$$

WHERE $P_n(\cos\theta)$ IS THE nTH-ORDER LEGENDRE POLYNOMIAL.

We can write the wave function of a symmetric top as:

$$|JKM\rangle \;\; = \;\; \psi_{JKM}(\phi,\theta,\chi)$$

$$= \;\; (-1)^{M-K}\left(\frac{2J+1}{8\pi^2}\right)^{\frac{1}{2}} D^J_{-M-K}(\phi,\theta,\chi)$$

$$= \;\; \left(\frac{2J+1}{8\pi^2}\right)^{\frac{1}{2}} D^{J*}_{MK}(\phi,\theta,\chi). \qquad (3.116)$$

By inserting equation 3.116 into equation 3.114, we obtain

$$P_{JKM}(\theta) \;\; = \;\; \int_0^{2\pi}\mathrm{d}\phi \int_0^{2\pi}\mathrm{d}\chi \left(\frac{2J+1}{8\pi^2}\right) D^J_{MK}(\phi,\theta,\chi)(-1)^{M-K}D^J_{-M-K}(\phi,\theta,\chi)$$

$$= \;\; \frac{2J+1}{8\pi^2}(-1)^{M-K}\int_0^{2\pi}\mathrm{d}\phi \int_0^{2\pi}\mathrm{d}\chi \sum_{J_3}(2J_3+1)D^{J_3}_{M_3'M_3}(\phi,\theta,\chi)$$

$$\times \begin{pmatrix} J & J & J_3 \\ M & -M & M_3 \end{pmatrix} \begin{pmatrix} J & J & J_3 \\ K & -K & M_3' \end{pmatrix}, \qquad (3.117)$$

where we have used the Clebsch-Gordan series. Using the properties of 3-j symbols, we can simplify equation 3.117:

$$
\begin{aligned}
P_{JKM}(\theta) \;=\;& \frac{2J+1}{8\pi^2}(-1)^{M-K}\int_0^{2\pi}\mathrm{d}\phi\int_0^{2\pi}\mathrm{d}\chi \\[2ex]
& \times\sum_{J_3}(2J_3+1)\begin{pmatrix} J & J & J_3 \\ M & -M & 0 \end{pmatrix}\begin{pmatrix} J & J & J_3 \\ K & -K & 0 \end{pmatrix}D_{00}^{J_3}(\phi,\theta,\chi) \\[3ex]
\;=\;& \frac{2J+1}{8\pi^2}(-1)^{M-K}\int_0^{2\pi}\mathrm{d}\phi\int_0^{2\pi}\mathrm{d}\chi \\[2ex]
& \times\sum_{J_3}(2J_3+1)\begin{pmatrix} J & J & J_3 \\ M & -M & 0 \end{pmatrix}\begin{pmatrix} J & J & J_3 \\ K & -K & 0 \end{pmatrix}P_{J_3}(\cos\theta) \\[3ex]
\;=\;& \frac{2J+1}{2}(-1)^{M-K}\sum_{J_3}(2J_3+1)\begin{pmatrix} J & J & J_3 \\ M & -M & 0 \end{pmatrix}\begin{pmatrix} J & J & J_3 \\ K & -K & 0 \end{pmatrix}P_{J_3}(\cos\theta),
\end{aligned}
$$

$$(3.118)$$

where we have used $D_{00}^{J_3}(\phi,\theta,\chi) = P_{J_3}(\cos\theta)$.

Calling $J_3 = n$, and considering that the sum goes from $n = 0$ to $n = 2J$ because it covers all possible values of J_3, we obtain the desired result, equation 3.115.

B.

FIND THE ORIENTATION OF THIS SYSTEM BY EVALUATING THE EXPECTATION VALUE OF $\cos\theta$. SHOW THAT

$$
\langle\cos\theta\rangle = \frac{MK}{J(J+1)}.
$$

$$(3.119)$$

$$
\begin{aligned}
\langle\cos\theta\rangle \;=\;& \int_0^{\pi} P_{JKM}(\theta)\cos\theta\sin\theta\,\mathrm{d}\theta \\[3ex]
\;=\;& \int_0^{\pi}\sin\theta\,\mathrm{d}\theta\,(-1)^{M-K}\,\frac{2J+1}{2}\sum_{n=0}^{2J}(2n+1)\begin{pmatrix} J & J & n \\ M & -M & 0 \end{pmatrix} \\[3ex]
& \times\begin{pmatrix} J & J & n \\ K & -K & 0 \end{pmatrix}P_n(\cos\theta)P_1(\cos\theta)
\end{aligned}
$$

$$= (-1)^{M-K} \frac{2J+1}{2} \sum_{n=0}^{2J} (2n+1) \begin{pmatrix} J & J & n \\ M & -M & 0 \end{pmatrix} \begin{pmatrix} J & J & n \\ K & -K & 0 \end{pmatrix} \frac{2}{2n+1} \delta_{n1}, \quad (3.120)$$

where we have used the facts that $\cos\theta = P_1(\cos\theta)$ and that

$$\int_0^\pi P_{\ell'}(\cos\theta) \, P_\ell(\theta) \sin\theta \, d\theta = \frac{2}{2\ell+1} \delta_{\ell\ell'}.$$

Because δ_{n1} cancels all the terms of the sum but the one for $n = 1$, equation 3.120 can be further simplified to give

$$\langle\cos\theta\rangle = (-1)^{M-K} (2J+1) \begin{pmatrix} J & J & 1 \\ M & -M & 0 \end{pmatrix} \begin{pmatrix} J & J & 1 \\ K & -K & 0 \end{pmatrix}$$

$$= (-1)^{M-K} (2J+1)(-1)^{J-M} \frac{2M}{[(2J+2)(2J+1)2J]^{\frac{1}{2}}} (-1)^{J-K} \frac{2K}{[(2J+2)(2J+1)2J]^{\frac{1}{2}}}$$

$$= \frac{MK}{J(J+1)}. \qquad (3.121)$$

C.

IN SECOND-ORDER PERTURBATION THEORY, THE MOLECULAR WAVE FUNCTION IS CHANGED BY THE PRESENCE OF A STATIC ELECTRIC FIELD, AND THE RESULTING ENERGY CHANGE IS GIVEN BY

$$\Delta E^{(2)} = \sum_{J'K'M'} \frac{\langle JKM| H'|J'K'M'\rangle \langle J'K'M'| H'|JKM\rangle}{E_{JKM} - E_{J'K'M'}}, \qquad (3.122)$$

WHERE THE SUMMATION IS OVER ALL $J'K'M' \neq JKM$. IN CHAPTER 5, THE ENERGY IS SHOWN TO BE GIVEN BY

$$E_{JKM} = BJ(J+1) + (C-B)K^2, \qquad (3.123)$$

WHERE B AND C ARE ROTATIONAL CONSTANTS. SHOW THAT THE SECOND-ORDER ENERGY IS AFFECTED ONLY BY THE TWO NEIGHBORING STATES $J' = J+1$ AND $J' = J-1$ WITH $K' = K$ AND $M' = M$, AND CALCULATE THE SUM OF THEIR EFFECTS.

To see which states affect the second-order energy, $\Delta E^{(2)}$, let us examine the matrix element

$$\langle JKM| H'|J'K'M'\rangle .$$

Only when this element is nonzero will $|J'K'M'\rangle$ affect the energy of $|JKM\rangle$ in the second-order perturbation theory analysis. This matrix element is given by

$$\langle JKM| H'|J'K'M'\rangle = -\mu_z E_Z \langle JKM|\cos\theta|J'K'M'\rangle$$

$$= -\mu_z E_Z \frac{\sqrt{(2J+1)(2J'+1)}}{8\pi^2} \int d\Omega D_{MK}^J(R) D_{00}^1(R) (-1)^{M'-K'} D_{-M'-K'}^{J'}(R)$$

$$= -\mu_z E_Z \sqrt{(2J+1)(2J'+1)}(-1)^{M'-K'}$$

$$\times \begin{pmatrix} J & J' & 1 \\ M & -M' & 0 \end{pmatrix} \begin{pmatrix} J & J' & 1 \\ K & -K' & 0 \end{pmatrix},$$ (3.124)

where we used equation Z-3.118. In evaluating equation 3.124, we realize that in order for $\langle JKM | H' | J'K'M' \rangle$ to be nonzero, the following conditions must hold:

$$M = M', \quad K = K', \quad J' = J \pm 1.$$

It follows that the second-order energy is affected only by the two neighboring states, $J' = J+1$ and $J' = J-1$, with $K' = K$ and $M' = M$.

Substituting equation 3.124 into equation 3.122, we obtain

$$\Delta E^{(2)} = \mu_z^2 E_Z^2 \sum_{J'K'M'} \frac{(2J+1)(2J'+1)(-1)^{M'-K'}(-1)^{M-K}}{E_{JKM} - E_{J'K'M'}}$$

$$\times \begin{pmatrix} J & J' & 1 \\ M & -M' & 0 \end{pmatrix} \begin{pmatrix} J & J' & 1 \\ K & -K' & 0 \end{pmatrix} \begin{pmatrix} J' & J & 1 \\ M' & -M & 0 \end{pmatrix} \begin{pmatrix} J' & J & 1 \\ K' & -K & 0 \end{pmatrix}.$$ (3.125)

The sums over M' and K' collapse into a single term with $M = M'$ and $K = K'$. Using the permutation properties of the 3-j symbols, we can simplify equation 3.125 to obtain

$$\Delta E^{(2)} = \mu_z^2 E_Z^2 \sum_{J'} \frac{(2J+1)(2J'+1)}{E_{JKM} - E_{J'K'M'}} \begin{pmatrix} J & 1 & J' \\ M & 0 & -M \end{pmatrix}^2 \begin{pmatrix} J & 1 & J' \\ K & 0 & -K \end{pmatrix}^2$$

$$= \mu_z^2 E_Z^2 \sum_{J' \neq J} \frac{(2J+1)(2J'+1)}{E_{JKM} - E_{J'K'M'}} (2J'+1)^{-2} \langle JM, 10 | J'M \rangle^2 \langle JK, 10 | J'K \rangle^2.$$ (3.126)

Substituting into equation 3.126 the energy eigenvalues for a symmetric top, we obtain

$$\Delta E^{(2)} = \mu_z^2 E_Z^2 \sum_{J'=J\pm 1} \langle JM, 10 | J'M \rangle^2 \langle JK, 10 | J'K \rangle^2 \frac{(2J+1)}{(2J'+1)}$$

$$\times [BJ(J+1) + (C-B)K^2 - BJ'(J'+1) - (C-B)K^2]^{-1}$$

$$= \mu_z^2 E_Z^2 \left[\frac{(2J+1)(J-M+1)(J+M+1)(J-K+1)(J+K+1)}{(2J+3)B[J(J+1) - (J+1)(J+2)](2J+1)^2(J+1)^2} \right.$$

$$\left. + \frac{(2J+1)(J-M)(J+M)(J-K)(J+K)}{(2J-1)B[J(J+1) - (J-1)J]J^2(2J+1)^2} \right]$$

$$= \frac{\mu_z^2 E_Z^2}{2B} \left[\frac{-[(J+1)^2 - M^2][(J+1)^2 - K^2]}{(2J+3)(2J+1)(J+1)^3} + \frac{(J^2 - M^2)(J^2 - K^2)}{(2J+1)J^3(2J-1)} \right].$$ (3.127)

For $K = 0$ we obtain the second-order Stark shift for a linear molecule in the ground vibrational level, or for a $^1\Sigma^+$ or $^1\Sigma^-$ diatomic molecule. Setting K=0 in equation 3.127 gives

$$\Delta E^{(2)} = \frac{\mu_z^2 E_Z^2}{2B} \left[\frac{J(J+1) - 3M^2}{J(J+1)(2J-1)(2J+3)} \right]. \tag{3.128}$$

For the special case of the $J = 0$ level, only the $J' = 1$ level can interact with it. Setting J=0 and M=0 in equation 3.128 gives

$$\Delta E^{(2)}(J=0) = \frac{-\mu_z^2 E_Z^2}{6B}. \tag{3.129}$$

APPLICATION 8
POLARIZED RESONANCE FLUORESCENCE AND POLARIZED RAMAN SCATTERING: CLASSICAL EXPRESSION

LET THE INCIDENT LIGHT BE LINEARLY POLARIZED WITH ITS ELECTRIC VECTOR $\mathbf{E}$ POINTING ALONG $\hat{\mathbf{F}}$ IN THE X, Y, OR Z DIRECTION OF THE LABORATORY-FIXED FRAME. THE LIGHT EXCITES AN ABSORPTION OSCILLATOR μ POINTING ALONG THE $\hat{\mathbf{g}}$ AXIS IN THE MOLECULAR FRAME. WE CHOOSE $\hat{\mathbf{g}}$ TO BE A UNIT VECTOR ALONG x, y, OR z AND DENOTE THE UNIT VECTORS PERPENDICULAR TO $\hat{\mathbf{g}}$ BY $\hat{\mathbf{g}}'$ AND $\hat{\mathbf{g}}''$. THE PROBABILITY OF EXCITATION IS PROPORTIONAL TO $\left|\mathbf{E}_F \cdot \mu_g\right|^2 = \Phi_{Fg}^2$, WHERE Φ_{Fg} IS THE DIRECTION COSINE MATRIX ELEMENT RELATING THE TWO UNIT VECTORS $\hat{\mathbf{F}}$ AND $\hat{\mathbf{g}}$, WHICH HAVE AS THEIR COMMON ORIGIN THE CENTER OF MASS OF THE MOLECULE.

LET THE EMISSION OSCILLATOR LIE ALONG $\hat{\mathbf{h}}$, WHICH DOES NOT NECESSARILY COINCIDE WITH $\hat{\mathbf{g}}$. THE PROBABILITY OF EMISSION WITH LIGHT PLANE-POLARIZED ALONG $\hat{\mathbf{H}}$ IS PROPORTIONAL TO $\left|\mathbf{E}_H \cdot \mu_h\right|^2 = \Phi_{Hh}^2$. IT IS TRADITIONAL TO CALL $I_\parallel$ THE INTENSITY OF THE LIGHT EMITTED PLANE-POLARIZED IN THE SAME DIRECTION AS THE ELECTRIC VECTOR OF THE INCIDENT LIGHT AND $I_\perp$ THE INTENSITY OF THE LIGHT EMITTED PLANE-POLARIZED PERPENDICULAR TO THE DIRECTION OF THE ELECTRIC VECTOR OF THE INCIDENT LIGHT,

$$I_\parallel = A\overline{\Phi_{Fg}^2 \Phi_{Fh}^2}, \quad \text{and} \quad I_\perp = A\overline{\Phi_{Fg}^2 \Phi_{F'h}^2}, \tag{3.130}$$

WHERE $\hat{\mathbf{F}}'$ IS PERPENDICULAR TO $\hat{\mathbf{F}}$, THE BARS REPRESENT AN ENSEMBLE AVERAGE OVER ALL INITIAL ORIENTATIONS OF THE MOLECULES, AND A IS A PROPORTIONALITY CONSTANT.

WE DEFINE THE DEGREE OF POLARIZATION P AND POLARIZATION ANISOTROPY R BY

$$P = \frac{I_\parallel - I_\perp}{I_\parallel + I_\perp} \quad \text{and} \quad R = \frac{I_\parallel - I_\perp}{I_\parallel + 2I_\perp}. \tag{3.131}$$

BY EXPANDING $\hat{\mathbf{h}}$ IN TERMS OF THE COMPONENTS OF THE $\hat{\mathbf{g}}$ FRAME,

$$\Phi_{Fh} = \sum_{g_1} \Phi_{Fg_1} \Phi_{g_1 h}, \tag{3.132}$$

WE OBTAIN THE EXPRESSIONS

$$I_\parallel = A \sum_{g_1} \sum_{g_2} \overline{\Phi_{Fg}^2 \Phi_{Fg_1} \Phi_{Fg_2}} \Phi_{g_1 h} \Phi_{g_2 h} \tag{3.133}$$

AND

$$I_\perp = A \sum_{g_1} \sum_{g_2} \overline{\Phi_{Fg}^2 \Phi_{F'g_1} \Phi_{F'g_2}} \Phi_{g_1 h} \Phi_{g_2 h}. \tag{3.134}$$

WE LIST BELOW THE ONLY NONVANISHING PRODUCTS OF FOUR DIRECTION COSINE MATRIX ELEMENTS AVERAGED OVER ALL MOLECULAR ORIENTATIONS:

$$\overline{\Phi_{Ai}^4} = \tfrac{1}{5}, \qquad \overline{\Phi_{Ai}^2 \Phi_{Aj}^2} = \tfrac{1}{15}, \qquad \overline{\Phi_{Ai}^2 \Phi_{Bi}^2} = \tfrac{1}{15},$$

$$\overline{\Phi_{Ai}^2 \Phi_{Bj}^2} = \tfrac{2}{15}, \qquad \overline{\Phi_{Ai} \Phi_{Aj} \Phi_{Bi} \Phi_{Bj}} = -\tfrac{1}{30}. \tag{3.135}$$

A.

SHOW THAT

$$I_\| = \frac{A}{15}(2\,\cos^2\gamma + 1) \quad \text{and} \quad I_\perp = \frac{A}{15}(2 - \cos^2\gamma),$$ (3.136)

FROM WHICH IT FOLLOWS THAT

$$P = \frac{3\cos^2\gamma - 1}{\cos^2\gamma + 3}$$ (3.137)

AND

$$R = \frac{2}{5}P_2(\cos\gamma),$$ (3.138)

WHERE γ IS THE ANGLE BETWEEN THE ABSORPTION AND EMISSION OSCILLATORS. *Hint*: MAKE USE OF THE IDENTITY

$$\Phi_{gh}^2 + \Phi_{g'h}^2 + \Phi_{g''h}^2 = 1.$$ (3.139)

Let us start by evaluating $I_\|$:

$$I_\| = A\sum_{g_1}\sum_{g_2}\overline{\Phi_{Fg}^2\Phi_{Fg_1}\Phi_{Fg_2}}\Phi_{g_1h}\Phi_{g_2h}$$

$$= \sum_{g_2}A\left[\overline{\Phi_{Fg}^2\Phi_{Fg}\Phi_{Fg_2}}\Phi_{gh}\Phi_{g_2h} + \overline{\Phi_{Fg}^2\Phi_{Fg'}\Phi_{Fg_2}}\Phi_{g'h}\Phi_{g_2h} + \overline{\Phi_{Fg}^2\Phi_{Fg''}\Phi_{Fg_2}}\Phi_{g''h}\Phi_{g_2h}\right]$$

$$= A\left[\overline{\Phi_{Fg}^2\Phi_{Fg}\Phi_{Fg}}\Phi_{gh}\Phi_{gh} + \overline{\Phi_{Fg}^2\Phi_{Fg}\Phi_{Fg'}}\Phi_{gh}\Phi_{g'h} + \overline{\Phi_{Fg}^2\Phi_{Fg}\Phi_{Fg''}}\Phi_{gh}\Phi_{g''h}\right.$$

$$+ \overline{\Phi_{Fg}^2\Phi_{Fg'}\Phi_{Fg}}\Phi_{g'h}\Phi_{gh} + \overline{\Phi_{Fg}^2\Phi_{Fg'}\Phi_{Fg'}}\Phi_{g'h}\Phi_{g'h} + \overline{\Phi_{Fg}^2\Phi_{Fg'}\Phi_{Fg''}}\Phi_{g'h}\Phi_{g''h}$$

$$\left. + \overline{\Phi_{Fg}^2\Phi_{Fg''}\Phi_{Fg}}\Phi_{g''h}\Phi_{gh} + \overline{\Phi_{Fg}^2\Phi_{Fg''}\Phi_{Fg'}}\Phi_{g''h}\Phi_{g'h} + \overline{\Phi_{Fg}^2\Phi_{Fg''}\Phi_{Fg''}}\Phi_{g''h}\Phi_{g''h}\right]$$

$$= A\left[\frac{1}{5}\Phi_{gh}^2 + \frac{1}{15}\Phi_{g'h}^2 + \frac{1}{15}\Phi_{g''h}^2\right]$$

$$= A\left[\frac{2}{15}\Phi_{gh}^2 + \frac{1}{15}\left(\Phi_{gh}^2 + \Phi_{g'h}^2 + \Phi_{g''h}^2\right)\right]$$

$$= \frac{A}{15}\left(2\cos^2\gamma + 1\right),$$ (3.140)

where we used equation 3.139 and the definition

$$\Phi_{gh}^2 = \cos^2 \gamma. \tag{3.141}$$

Using the same procedure, we can evaluate $I_\perp$:

$$
\begin{aligned}
I_\perp &= A\left[\overline{\Phi_{Fg}^2 \Phi_{Hg}^2}\Phi_{gh}^2 + \overline{\Phi_{Fg}^2 \Phi_{Hg} \Phi_{Hg'}}\Phi_{gh}\Phi_{g'h} + \overline{\Phi_{Fg}^2 \Phi_{Hg} \Phi_{Hg''}}\Phi_{gh}\Phi_{g''h}\right.\\[2mm]
&\quad + \overline{\Phi_{Fg}^2 \Phi_{Hg'} \Phi_{Hg}}\Phi_{g'h}\Phi_{gh} + \overline{\Phi_{Fg}^2 \Phi_{Hg'}^2}\Phi_{g'h}^2 + \overline{\Phi_{Fg}^2 \Phi_{Hg'} \Phi_{Hg''}}\Phi_{g'h}\Phi_{g''h}\\[2mm]
&\quad \left.+ \overline{\Phi_{Fg}^2 \Phi_{Hg''} \Phi_{Hg}}\Phi_{g''h}\Phi_{gh} + \overline{\Phi_{Fg}^2 \Phi_{Hg''} \Phi_{Hg'}}\Phi_{g''h}\Phi_{g'h} + \overline{\Phi_{Fg}^2 \Phi_{Hg''}^2}\Phi_{g''h}^2\right]\\[3mm]
&= A\left[\frac{1}{15}\Phi_{gh}^2 + \frac{2}{15}\Phi_{g'h}^2 + \frac{2}{15}\Phi_{g''h}^2\right]\\[3mm]
&= A\left[\frac{2}{15}\left(\Phi_{gh}^2 + \Phi_{g'h}^2 + \Phi_{g''h}^2\right) - \frac{1}{15}\Phi_{gh}^2\right]\\[3mm]
&= \frac{A}{15}\left(2 - \cos^2 \gamma\right). \tag{3.142}
\end{aligned}
$$

Substituting $I_\parallel$ and $I_\perp$ into equation 3.131, we obtain

$$
\begin{aligned}
P &= \frac{2\cos^2 \gamma + 1 - 2 + \cos^2 \gamma}{2\cos^2 \gamma + 1 + 2 - \cos^2 \gamma}\\[3mm]
&= \frac{3\cos^2 \gamma - 1}{\cos^2 \gamma + 3}, \tag{3.143}
\end{aligned}
$$

and

$$
\begin{aligned}
R &= \frac{2\cos^2 \gamma + 1 - 2 + \cos^2 \gamma}{2\cos^2 \gamma + 1 + 4 - 2\cos^2 \gamma}\\[3mm]
&= \frac{1}{5}\left(3\cos^2 \gamma - 1\right)\\[3mm]
&= \frac{2}{5}P_2(\cos \gamma). \tag{3.144}
\end{aligned}
$$

B.

DERIVE EQUATION 3.135. *Hint*: IT IS READILY VERIFIED THAT

$$\overline{\Phi_{Fg}^4} = \frac{1}{4\pi} \int_0^{2\pi} \int_0^{\pi} \cos\theta^4 \sin\theta d\theta d\phi = \frac{1}{5}, \tag{3.145}$$

WHERE θ IS THE ANGLE BETWEEN $\hat{\mathbf{F}}$ AND $\hat{\mathbf{g}}$. CONSIDER THE TWO EQUATIONS

$$\sum_g \Phi_{Fg}^2 = 1 \qquad \text{and} \qquad \sum_g \Phi_{Fg}\Phi_{F'g} = 0. \tag{3.146}$$

SQUARE EACH EQUATION AND CARRY OUT THE AVERAGE OVER ALL MOLECULAR ORIENTATIONS. COMPLETE THIS PROBLEM BY CONSIDERING THE IDENTITY $\left(\sum_F \Phi_{Fg}^2\right)\left(\sum_{F'} \Phi_{F'g'}^2\right) = 1$.

Because the direction cosine matrix is a real unitary matrix, it has the following properties (equation Z-3.38):

$$\sum_i \Phi_{Ai}^2 = 1, \tag{3.147}$$

$$\sum_i \Phi_{Ai}\Phi_{Bi} = 0 \quad \text{for } A \neq B, \tag{3.148}$$

$$\sum_A \Phi_{Ai}^2 = 1, \tag{3.149}$$

and

$$\sum_A \Phi_{Ai}\Phi_{Aj} = 0 \quad \text{for } i \neq j. \tag{3.150}$$

The average of a function $\mathcal{F}$ over all molecular orientations is given by

$$\overline{\mathcal{F}} = \frac{1}{8\pi^2} \int_0^{2\pi} \int_0^{2\pi} \int_0^{\pi} \mathcal{F} d\phi d\chi \sin\theta d\theta. \tag{3.151}$$

It is clear from equation 3.151 that

$$\overline{\mathcal{F} + \mathcal{J}} = \overline{\mathcal{F}} + \overline{\mathcal{J}}. \tag{3.152}$$

Without any loss of generality, we can set θ to be the angle between $\hat{\mathbf{A}}$ ($\hat{\mathbf{F}}$, $\hat{\mathbf{F}}'$ or $\hat{\mathbf{F}}''$) in the laboratory frame and $\hat{\mathbf{i}}$ ($\hat{\mathbf{g}}$, $\hat{\mathbf{g}}'$ or $\hat{\mathbf{g}}''$) in the molecular frame, so that

$$\Phi_{Ai} = \cos\theta. \tag{3.153}$$

The orientation averages of the products of direction cosines are obtained as follows:

i) $\overline{\Phi_{Ai}^4}$

$$\overline{\Phi_{Ai}^4} \;=\; \frac{1}{8\pi^2} \int_0^{2\pi} \int_0^{2\pi} \int_0^{\pi} \Phi_{Ai}^4 \mathrm{d}\phi \mathrm{d}\chi \sin\theta \mathrm{d}\theta$$

$$=\; \frac{1}{2} \int_0^{\pi} \cos^4\theta \sin\theta \mathrm{d}\theta$$

$$=\; \frac{1}{5}. \tag{3.154}$$

ii) $\overline{\Phi_{Ai}^2 \Phi_{Aj}^2}$

From equation 3.147, we know that

$$1 \;=\; \overline{\left(\sum_i \Phi_{Ai}^2 \right)^2}$$

$$=\; \overline{\Phi_{Ag}^4} + \overline{\Phi_{Ag'}^4} + \overline{\Phi_{Ag''}^4} + 2 \left[\overline{\Phi_{Ag}^2 \Phi_{Ag'}^2} + \overline{\Phi_{Ag}^2 \Phi_{Ag''}^2} + \overline{\Phi_{Ag''}^2 \Phi_{Ag'}^2} \right]. \tag{3.155}$$

Because we are taking the average over the randomly oriented molecular ensemble, all $\overline{\Phi_{Ai}^2 \Phi_{Aj}^2}$ with $i \neq j$ are equal. We therefore get

$$1 = \frac{1}{5} + \frac{1}{5} + \frac{1}{5} + 6 \left(\overline{\Phi_{Ai}^2 \Phi_{Aj}^2} \right), \tag{3.156}$$

from which we conclude that

$$\overline{\Phi_{Ai}^2 \Phi_{Aj}^2} \;=\; \frac{1}{15}. \tag{3.157}$$

iii) $\overline{\Phi_{Ai}^2 \Phi_{Bi}^2}$

We repeat the same procedure as before, but start instead with equation 3.149

$$1 \;=\; \overline{\left(\sum_A \Phi_{Ai}^2 \right)^2}$$

$$=\; \overline{\Phi_{Fi}^4} + \overline{\Phi_{F'i}^4} + \overline{\Phi_{F''i}^4} + 2 \left[\overline{\Phi_{Fi}^2 \Phi_{F'i}^2} + \overline{\Phi_{Fi}^2 \Phi_{F''i}^2} + \overline{\Phi_{F'i}^2 \Phi_{F''i}^2} \right]. \tag{3.158}$$

With the same arguments used in part (*ii*), we note that $\overline{\Phi_{Ai}^2 \Phi_{Bi}^2}$ is the same for all $A \neq B$, so that

$$1 = \frac{3}{5} + 6 \left(\overline{\Phi_{Ai}^2 \Phi_{Bi}^2} \right), \tag{3.159}$$

from which we conclude that

$$\overline{\Phi^2_{Ai}\Phi^2_{Bi}} = \frac{1}{15}. \tag{3.160}$$

$iv)$ $\overline{\Phi^2_{Ai}\Phi^2_{Bj}}$

From equation 3.149, we know that

$$1 = \overline{\sum_A \Phi^2_{Ai} \sum_B \Phi^2_{Bj}}$$

$$= \overline{\left(\Phi^2_{Fi} + \Phi^2_{F'i} + \Phi^2_{F''i}\right)\left(\Phi^2_{Fj} + \Phi^2_{F'j} + \Phi^2_{F''j}\right)}$$

$$= \overline{\Phi^2_{Fi}\Phi^2_{Fj}} + \overline{\Phi^2_{F'i}\Phi^2_{F'j}} + \overline{\Phi^2_{F''i}\Phi^2_{F''j}} + \overline{\Phi^2_{Fi}\Phi^2_{F'j}} + \overline{\Phi^2_{Fi}\Phi^2_{F''j}}$$

$$+ \overline{\Phi^2_{F'i}\Phi^2_{Fj}} + \overline{\Phi^2_{F'i}\Phi^2_{F''j}} + \overline{\Phi^2_{F''i}\Phi^2_{Fj}} + \overline{\Phi^2_{F''i}\Phi^2_{F'j}}. \tag{3.161}$$

All the $\overline{\Phi^2_{Ai}\Phi^2_{Bj}}$ are equal because of symmetry. We therefore obtain

$$1 = \frac{3}{15} + 6\left(\overline{\Phi^2_{Ai}\Phi^2_{Bj}}\right), \tag{3.162}$$

from which we conclude that

$$\overline{\Phi^2_{Ai}\Phi^2_{Bj}} = \frac{2}{15}. \tag{3.163}$$

$v)$ $\overline{\Phi_{Ai}\Phi_{Aj}\Phi_{Bi}\Phi_{Bj}}$

From equation 3.148, we know that

$$0 = \overline{\left(\sum_i \Phi_{Ai}\Phi_{Bi}\right)^2}$$

$$= \overline{\left(\Phi_{Ag}\Phi_{Bg} + \Phi_{Ag'}\Phi_{Bg'} + \Phi_{Ag''}\Phi_{Bg''}\right)^2}$$

$$= \overline{\Phi^2_{Ag}\Phi^2_{Bg}} + \overline{\Phi^2_{Ag'}\Phi^2_{Bg'}} + \overline{\Phi^2_{Ag''}\Phi^2_{Bg''}}$$

$$2\left(\overline{\Phi_{Ag}\Phi_{Bg}\Phi_{Ag'}\Phi_{Bg'}} + \overline{\Phi_{Ag}\Phi_{Bg}\Phi_{Ag''}\Phi_{Bg''}} + \overline{\Phi_{Ag'}\Phi_{Bg'}\Phi_{Ag''}\Phi_{Bg''}}\right). \tag{3.164}$$

By symmetry, all of the $\overline{\Phi_{Ai}\Phi_{Bi}\Phi_{Aj}\Phi_{Bj}}$ terms with $i \neq j$ are equal. We therefore get

$$0 = 3\left(\overline{\Phi_{Ag}^2\Phi_{Bg}^2}\right) + 6\left(\overline{\Phi_{Ag}\Phi_{Bg}\Phi_{Ag'}\Phi_{Bg'}}\right)$$

$$= \frac{3}{15} + 6\left(\overline{\Phi_{Ai}\Phi_{Bi}\Phi_{Aj}\Phi_{Bj}}\right), \tag{3.165}$$

from which we conclude that

$$\overline{\Phi_{Ai}\Phi_{Bi}\Phi_{Aj}\Phi_{Bj}} = -\frac{1}{30}. \tag{3.166}$$

C.

OFTEN MOLECULAR RESONANCE FLUORESCENCE IS NOT RESOLVED. FIND AN EXPRESSION FOR THE AP-
PARENT POLARIZATION IF THE RESONANCE FLUORESCENCE PROCESS CONSISTS OF TWO TYPES OF EMIS-
SION, ONE OF TOTAL INTENSITY I_1 OF POLARIZATION P_1 AND THE OTHER OF TOTAL INTENSITY I_2 OF
POLARIZATION P_2. *Hint:* R IS LINEAR IN $\cos^2 \gamma$ WHILE P IS NOT.

The intensity of the perpendicular component is twice that of the parallel component. The total intensity
for each polarization is given by

$$I_1 = I_{1\|} + 2I_{1\perp}, \tag{3.167}$$

and

$$I_2 = I_{2\|} + 2I_{2\perp}. \tag{3.168}$$

We further define

$$I_{\|} = I_{1\|} + I_{2\|} \tag{3.169}$$

and

$$I_{\perp} = I_{1\perp} + I_{2\perp}. \tag{3.170}$$

Because R is linear in $\cos^2 \gamma$, it is convenient to calculate its apparent value, R_{app}, first and subsequently
convert it to the apparent polarization ratio, P_{app}. The polarization and anisotropy ratios for the two types
of emission are defined by

$$P_1 = \frac{I_{1\|} - I_{1\perp}}{I_{1\|} + I_{1\perp}}, \tag{3.171}$$

$$P_2 = \frac{I_{2\|} - I_{2\perp}}{I_{2\|} + I_{2\perp}}, \tag{3.172}$$

$$R_1 = \frac{I_{1\|} - I_{1\perp}}{I_1}, \tag{3.173}$$

and

$$R_2 = \frac{I_{2\parallel} - I_{2\perp}}{I_2}. \tag{3.174}$$

The apparent anisotropy is defined as

$$R_{\mathrm{app}} = \frac{I_\parallel - I_\perp}{I_1 + I_2} \tag{3.175}$$

$$= \frac{I_{1\parallel} + I_{2\parallel} - I_{1\perp} - I_{2\perp}}{I_1 + I_2}$$

$$= \frac{\left(\frac{I_{1\parallel} - I_{1\perp}}{I_1}\right) I_1 + \left(\frac{I_{2\parallel} - I_{2\perp}}{I_2}\right) I_2}{I_1 + I_2}$$

$$= c_1 R_1 + c_2 R_2, \tag{3.176}$$

where

$$c_i = \frac{I_i}{I_1 + I_2} \tag{3.177}$$

for $i = 1$ and 2. From equation Z-5.113 we can write

$$R_1 = \frac{2P_1}{3 - P_1} \quad \text{and} \quad R_2 = \frac{2P_2}{3 - P_2}, \tag{3.178}$$

which can be used to express R_{app} in equation 3.176 in terms of P_1 and P_2.

Another way of describing the polarization is to define the apparent polarization angle γ_{app} by the expression

$$R_{\mathrm{app}} = \frac{2}{5} P_2(\cos \gamma_{\mathrm{app}})$$

$$= \frac{1}{5}(3 \cos^2 \gamma_{\mathrm{app}} - 1). \tag{3.179}$$

From equation 3.144, we can write

$$R_1 = \frac{2}{5} P_2(\cos \gamma_1) = \frac{1}{5}(3 \cos^2 \gamma_1 - 1) \tag{3.180}$$

and

$$R_2 = \frac{2}{5} P_2(\cos \gamma_2) = \frac{1}{5}(3 \cos^2 \gamma_2 - 1). \tag{3.181}$$

Substituting equations 3.180 and 3.181 into equation 3.176, we obtain

$$R_{\mathrm{app}} = \frac{c_1}{5}(3 \cos^2 \gamma_1 - 1) + \frac{c_2}{5}(3 \cos^2 \gamma_2 - 1). \tag{3.182}$$

Comparing equations 3.179 and 3.182, we obtain

$$\cos^2 \gamma_{\mathrm{app}} \;=\; c_1 \cos^2 \gamma_1 + c_2 \cos^2 \gamma_2. \tag{3.183}$$

We define the apparent degree of polarization by

$$P_{\mathrm{app}} \;=\; \frac{I_{\parallel} - I_{\perp}}{I_{\parallel} + I_{\perp}}. \tag{3.184}$$

Comparing equations 3.175 and 3.184, we obtain the relation

$$R_{\mathrm{app}} \;=\; \frac{2 P_{\mathrm{app}}}{3 - P_{\mathrm{app}}} \tag{3.185}$$

and its inverse

$$P_{\mathrm{app}} \;=\; \frac{3 R_{\mathrm{app}}}{2 + R_{\mathrm{app}}}. \tag{3.186}$$

Substituting equation 3.179 into 3.186 gives the result

$$P_{\mathrm{app}} \;=\; \frac{\frac{6}{5} P_2(\cos \gamma_{\mathrm{app}})}{2 + \frac{2}{5} P_2(\cos \gamma_{\mathrm{app}})}$$

$$\;=\; \frac{3 \cos^2 \gamma_{\mathrm{app}} - 1}{\cos^2 \gamma_{\mathrm{app}} + 3}. \tag{3.187}$$

D.

OFF RESONANCE, WE MAY OBSERVE THE *Raman effect*. HERE THE ELECTRIC FIELD OF THE INCIDENT LIGHT $\mathbf{E}$ INDUCES A DIPOLE MOMENT $\boldsymbol{\mu}$ ACCORDING TO THE RELATION

$$\boldsymbol{\mu} = \underset{\sim}{\boldsymbol{\alpha}}\, \mathbf{E}, \tag{3.188}$$

WHERE $\underset{\sim}{\boldsymbol{\alpha}}$ IS THE *molecular polarizability tensor*.

THE DIRECTION COSINE MATRIX $\underset{\sim}{\boldsymbol{\Phi}}$ EXPRESSES THE TRANSFORMATION FROM THE SPACE-FIXED TO THE MOLECULE-FIXED FRAMES. WE HAVE

$$\boldsymbol{\mu}^{\mathrm{mol}} = \underset{\sim}{\boldsymbol{\Phi}}\, \boldsymbol{\mu}^{\mathrm{space}} \qquad \text{and} \qquad \mathbf{E}^{\mathrm{mol}} = \underset{\sim}{\boldsymbol{\Phi}}\, \mathbf{E}^{\mathrm{space}}. \tag{3.189}$$

THE POLARIZABILITY TENSOR IN THE MOLECULE-FIXED FRAME IS GIVEN BY

$$\boldsymbol{\mu}^{\mathrm{mol}} = \underset{\sim}{\boldsymbol{\alpha}}^{\mathrm{mol}} \mathbf{E}^{\mathrm{mol}}. \tag{3.190}$$

SHOW THAT

$$\underset{\sim}{\boldsymbol{\alpha}}^{\mathrm{space}} = \underset{\sim}{\boldsymbol{\Phi}}^{-1} \underset{\sim}{\boldsymbol{\alpha}}^{\mathrm{mol}} \underset{\sim}{\boldsymbol{\Phi}} \qquad \text{and} \qquad \underset{\sim}{\boldsymbol{\alpha}}^{\mathrm{mol}} = \underset{\sim}{\boldsymbol{\Phi}}\, \underset{\sim}{\boldsymbol{\alpha}}^{\mathrm{space}} \underset{\sim}{\boldsymbol{\Phi}}^{-1}. \tag{3.191}$$

The direction cosine matrix is real and unitary; that is,

$$\underset{\sim}{\Phi}\ \underset{\sim}{\Phi}^{-1} = \underset{\sim}{\Phi}^{-1}\ \underset{\sim}{\Phi} = \underset{\sim}{I}. \tag{3.192}$$

Multiplying equation 3.189 by $\underset{\sim}{\Phi}^{-1}$ and using equation 3.192, we obtain

$$\mu^{\text{space}} = \underset{\sim}{\Phi}^{-1}\mu^{\text{mol}} \qquad \text{and} \qquad \mathbf{E}^{\text{space}} = \underset{\sim}{\Phi}^{-1}\mathbf{E}^{\text{mol}}. \tag{3.193}$$

From equation 3.188, we know that

$$\mu^{\text{space}} = \underset{\sim}{\alpha}^{\text{space}}\ \mathbf{E}^{\text{space}}. \tag{3.194}$$

Inserting equation 3.193 into equation 3.194, we obtain

$$\underset{\sim}{\Phi}^{-1}\mu^{\text{mol}} = \underset{\sim}{\alpha}^{\text{space}}\underset{\sim}{\Phi}^{-1}\mathbf{E}^{\text{mol}}, \tag{3.195}$$

and using equation 3.192,

$$\mu^{\text{mol}} = \underset{\sim}{\Phi}\ \underset{\sim}{\alpha}^{\text{space}}\underset{\sim}{\Phi}^{-1}\mathbf{E}^{\text{mol}}. \tag{3.196}$$

Comparing equations 3.196 and 3.190, we see that

$$\underset{\sim}{\alpha}^{\text{mol}} = \underset{\sim}{\Phi}\ \underset{\sim}{\alpha}^{\text{space}}\underset{\sim}{\Phi}^{-1}. \tag{3.197}$$

Multiplying equation 3.197 by $\underset{\sim}{\Phi}^{-1}$ and $\underset{\sim}{\Phi}$ on both sides, we obtain

$$\underset{\sim}{\Phi}^{-1}\underset{\sim}{\alpha}^{\text{mol}}\underset{\sim}{\Phi} = \underset{\sim}{\Phi}^{-1}\underset{\sim}{\Phi}\ \underset{\sim}{\alpha}^{\text{space}}\underset{\sim}{\Phi}^{-1}\underset{\sim}{\Phi}$$

$$= \underset{\sim}{\alpha}^{\text{space}}. \tag{3.198}$$

E.

We define the quantities

$$a = \frac{1}{3}(\alpha_{xx} + \alpha_{yy} + \alpha_{zz}) \tag{3.199}$$

AND

$$\gamma^2 = \frac{1}{2}[(\alpha_{xx} - \alpha_{yy})^2 + (\alpha_{yy} - \alpha_{zz})^2 + (\alpha_{zz} - \alpha_{xx})^2]. \tag{3.200}$$

Show that the depolarization ratio

$$\rho = \frac{I_\perp}{I_\parallel} \tag{3.201}$$

is given by

$$\rho = \frac{3\gamma^2}{45a^2 + 4\gamma^2}. \tag{3.202}$$

Hint: USE EQUATIONS Z-3.8.22 AND 3.198 TO FIND EXPRESSIONS FOR α_{ZZ} AND α_{XZ}. SQUARE THESE EXPRESSIONS AND USE EQUATION 3.135 TO CARRY OUT THE AVERAGES OVER ALL MOLECULAR ORIENTATIONS.

The direction cosine matrix has the form

$$\underset{\sim}{\Phi} = \begin{pmatrix} \Phi_{xX} & \Phi_{xY} & \Phi_{xZ} \\ \Phi_{yX} & \Phi_{yY} & \Phi_{yZ} \\ \Phi_{zX} & \Phi_{zY} & \Phi_{zZ} \end{pmatrix}. \tag{3.203}$$

Because $\underset{\sim}{\Phi}$ is a real unitary matrix, the inverse of the direction cosine matrix, $\underset{\sim}{\Phi}^{-1}$, is just the transpose of $\underset{\sim}{\Phi}$. Inserting equations Z-3.8.22 and 3.203 into equation 3.198, we obtain

$$\underset{\sim}{\alpha}^{\text{space}} = \begin{pmatrix} \Phi_{xX} & \Phi_{yX} & \Phi_{zX} \\ \Phi_{xY} & \Phi_{yY} & \Phi_{zY} \\ \Phi_{xZ} & \Phi_{yZ} & \Phi_{zZ} \end{pmatrix} \begin{pmatrix} \alpha_{xx} & 0 & 0 \\ 0 & \alpha_{yy} & 0 \\ 0 & 0 & \alpha_{zz} \end{pmatrix} \begin{pmatrix} \Phi_{xX} & \Phi_{xY} & \Phi_{xZ} \\ \Phi_{yX} & \Phi_{yY} & \Phi_{yZ} \\ \Phi_{zX} & \Phi_{zY} & \Phi_{zZ} \end{pmatrix}. \tag{3.204}$$

Because $I_\parallel$ is proportional to $\overline{\alpha_{ZZ}^2}$ and $I_\perp$ is proportional to $\overline{\alpha_{XZ}^2}$, we need to calculate only those two matrix elements. The results are

$$\overline{\alpha_{ZZ}^2} = \overline{\left(\Phi_{xZ}\alpha_{xx}\Phi_{xZ} + \Phi_{yZ}\alpha_{yy}\Phi_{yZ} + \Phi_{zZ}\alpha_{zz}\Phi_{zZ} \right)^2}$$

$$= \overline{\left(\alpha_{xx}\Phi_{xZ}^2 + \alpha_{yy}\Phi_{yZ}^2 + \alpha_{zz}\Phi_{zZ}^2 \right)^2}$$

$$= \alpha_{xx}^2 \overline{\Phi_{xZ}^4} + \alpha_{yy}^2 \overline{\Phi_{yZ}^4} + \alpha_{zz}^2 \overline{\Phi_{zZ}^4}$$

$$\quad + 2\left(\alpha_{xx}\alpha_{yy}\overline{\Phi_{xZ}^2\Phi_{yZ}^2} + \alpha_{xx}\alpha_{zz}\overline{\Phi_{xZ}^2\Phi_{zZ}^2} + \alpha_{yy}\alpha_{zz}\overline{\Phi_{yZ}^2\Phi_{zZ}^2} \right)$$

$$= \frac{1}{5}\left(\alpha_{xx}^2 + \alpha_{yy}^2 + \alpha_{zz}^2 \right) + \frac{2}{15}\left(\alpha_{xx}\alpha_{yy} + \alpha_{xx}\alpha_{zz} + \alpha_{yy}\alpha_{zz} \right)$$

$$= \frac{1}{45}\left\{ 5\left(\alpha_{xx} + \alpha_{yy} + \alpha_{zz} \right)^2 + 2\left[(\alpha_{xx} - \alpha_{yy})^2 + (\alpha_{yy} - \alpha_{zz})^2 + (\alpha_{zz} - \alpha_{xx})^2 \right] \right\}$$

$$= \frac{45a^2 + 4\gamma^2}{45}, \tag{3.205}$$

and

$$
\begin{aligned}
\overline{\alpha_{XZ}^2} &= \overline{\left(\Phi_{xX}\alpha_{xx}\Phi_{xZ} + \Phi_{yX}\alpha_{yy}\Phi_{yZ} + \Phi_{zX}\alpha_{zz}\Phi_{zZ}\right)^2} \\[2mm]
&= \alpha_{xx}^2\,\overline{\Phi_{xX}^2\Phi_{xZ}^2} + \alpha_{yy}^2\,\overline{\Phi_{yX}^2\Phi_{yZ}^2} + \alpha_{zz}^2\,\overline{\Phi_{zX}^2\Phi_{zZ}^2} \\[2mm]
&\quad + 2\left(\alpha_{xx}\alpha_{yy}\,\overline{\Phi_{xX}\Phi_{xZ}\Phi_{yX}\Phi_{yZ}} + \alpha_{xx}\alpha_{zz}\,\overline{\Phi_{xX}\Phi_{xZ}\Phi_{zX}\Phi_{zZ}} + \alpha_{yy}\alpha_{zz}\,\overline{\Phi_{yX}\Phi_{yZ}\Phi_{zX}\Phi_{zZ}}\right) \\[2mm]
&= \frac{1}{15}\left(\alpha_{xx}^2 + \alpha_{yy}^2 + \alpha_{zz}^2\right) - \frac{1}{15}\left(\alpha_{xx}\alpha_{yy} + \alpha_{xx}\alpha_{zz} + \alpha_{yy}\alpha_{zz}\right) \\[2mm]
&= \frac{1}{30}\left[(\alpha_{xx} - \alpha_{yy})^2 + (\alpha_{yy} - \alpha_{zz})^2 + (\alpha_{zz} - \alpha_{xx})^2\right] \\[2mm]
&= \frac{\gamma^2}{15}.
\end{aligned}
\tag{3.206}
$$

Using equations 3.205 and 3.206, we can now calculate the depolarization ratio

$$
\begin{aligned}
\rho &= \frac{I_\perp}{I_\parallel} \\[2mm]
&= \frac{\overline{\alpha_{XZ}^2}}{\overline{\alpha_{ZZ}^2}} \\[2mm]
&= \frac{3\gamma^2}{45a^2 + 4\gamma^2}.
\end{aligned}
\tag{3.207}
$$

F.

CALCULATE ρ FOR RAYLEIGH SCATTERING FROM CH_4.

The fact that CH_4 is a spherical top implies that

$$
\alpha_{xx} = \alpha_{yy} = \alpha_{zz} = \alpha.
\tag{3.208}
$$

Inserting equation 3.208 into equations 3.199 and 3.200, we obtain

$$
a = \alpha,
$$

and

$$\gamma = 0,$$

which can be inserted into equation 3.207 to give

$$\rho = 0.$$

APPLICATION 9
MAGNETIC DEPOLARIZATION OF RESONANCE FLUORESCENCE: ZEEMAN QUANTUM BEATS AND THE HANLE EFFECT

LET US CONSIDER A CASE IN WHICH A MOLECULAR SAMPLE IS LOCATED AT THE ORIGIN, A MAGNETIC FIELD $\mathbf{H}$ IS DIRECTED ALONG THE Z AXIS, AND AN INCIDENT PLANE-POLARIZED LIGHT BEAM PROPAGATES ALONG THE X AXIS TOWARD THE ORIGIN WITH ITS ELECTRIC VECTOR POINTING ALONG THE Y AXIS. THE RESULTING RESONANCE FLUORESCENCE IS DETECTED ALONG THE Y AXIS AND ANALYZED FOR ITS POLARIZATION ALONG THE X AXIS. IF AT A TIME $t = t_0$ A SHORT PULSE OF LIGHT IS INCIDENT ON THE SAMPLE, THEN AT A LATER TIME t THE RESONANCE FLUORESCENCE SIGNAL WILL BE GIVEN BY

$$I(H, t - t_0) = A\overline{\Phi_{Yg}^2 \Phi_{Xh}^2}(H, t - t_0)\, e^{-\frac{t-t_0}{\tau}}. \tag{3.209}$$

A.

SHOW THAT

$$I(H, t - t_0) = A'[1 - P\cos 2\omega_L(t - t_0)]\, e^{-\frac{t-t_0}{\tau}},$$

WHERE

$$P = \frac{I_\parallel - I_\perp}{I_\parallel + I_\perp}.$$

In the laboratory frame, the total angular momentum, $\mathbf{J}$, and the molecular-fixed axes, $\hat{\mathbf{g}}$ and $\hat{\mathbf{h}}$, precess about $\mathbf{H}$ with the Larmor frequency, ω_L. From the molecule-fixed point of view, the laboratory-fixed axes $\hat{\mathbf{X}}$ and $\hat{\mathbf{Y}}$ precess with the Larmor frequency in the opposite direction. From equation Z-3.9.2, we can write

$$\begin{pmatrix} \hat{\mathbf{X}}(\mathbf{t}) \\ \hat{\mathbf{Y}}(\mathbf{t}) \\ \hat{\mathbf{Z}}(\mathbf{t}) \end{pmatrix} = \begin{pmatrix} \cos\omega_L(t - t_0) & \sin\omega_L(t - t_0) & 0 \\ -\sin\omega_L(t - t_0) & \cos\omega_L(t - t_0) & 0 \\ 0 & 0 & 1 \end{pmatrix} \begin{pmatrix} \hat{\mathbf{X}}(\mathbf{t_0}) \\ \hat{\mathbf{Y}}(\mathbf{t_0}) \\ \hat{\mathbf{Z}}(\mathbf{t_0}) \end{pmatrix}, \tag{3.210}$$

from which we obtain

$$\hat{\mathbf{X}}(\mathbf{t}) = \hat{\mathbf{X}}(\mathbf{t_0})\cos\omega_L(\mathbf{t} - \mathbf{t_0}) + \hat{\mathbf{Y}}(\mathbf{t_0})\sin\omega_L(\mathbf{t} - \mathbf{t_0}). \tag{3.211}$$

If we adopt the molecule-fixed point of view, we can rewrite equation 3.209 as

$$I(H, t - t_0) = A\overline{\Phi_{Y(t_0)g}^2 \Phi_{X(t)h}^2}\, e^{-\frac{t-t_0}{\tau}}. \tag{3.212}$$

Substituting $X(t)$ into equation 3.212, we obtain

$$\begin{aligned} I(H, t - t_0) &= A\overline{\Phi_{Y(t_0)g}^2 \left[\Phi_{X(t_0)h}\cos\omega_L(t - t_0) + \Phi_{Y(t_0)h}\sin\omega_L(t - t_0)\right]^2}\, e^{-\frac{t-t_0}{\tau}} \\[2mm] &= A\Big[\overline{\Phi_{Y(t_0)g}^2 \Phi_{X(t_0)h}^2}\cos^2\omega_L(t - t_0) + \overline{\Phi_{Y(t_0)g}^2 \Phi_{Y(t_0)h}^2}\sin^2\omega_L(t - t_0) \\[2mm] &\quad + 2\,\overline{\Phi_{Y(t_0)g}^2 \Phi_{X(t_0)h}\Phi_{Y(t_0)h}}\cos\omega_L(t - t_0)\sin\omega_L(t - t_0)\Big]\, e^{-\frac{t-t_0}{\tau}}. \end{aligned} \tag{3.213}$$

From Application 8, equations Z-3.8.1 , Z-3.8.2, and Z-3.8.11, we recall that

$$I_\parallel = A'' \overline{\Phi_{Y(t_0)g}^2 \Phi_{Y(t_0)h}^2} \qquad (3.214)$$

and

$$I_\perp = A'' \overline{\Phi_{Y(t_0)g}^2 \Phi_{X(t_0)h}^2} \, , \qquad (3.215)$$

where A'' is a proportionality constant. We also know that

$$\overline{\Phi_{Y(t_0)g}^2 \Phi_{X(t_0)h} \Phi_{Y(t_0)h}} = 0. \qquad (3.216)$$

Using equations 3.214 through 3.216, equation 3.213 becomes

$$
\begin{aligned}
I(H, t - t_0) &= A''' \left[I_\perp \cos^2 \omega_L(t - t_0) + I_\parallel \sin^2 \omega_L(t - t_0) \right] e^{-\frac{t-t_0}{\tau}} \\[2mm]
&= A''' \left[I_\perp \left(\frac{1 + \cos 2\omega_L(t - t_0)}{2} \right) + I_\parallel \left(\frac{1 - \cos 2\omega_L(t - t_0)}{2} \right) \right] e^{-\frac{t-t_0}{\tau}} \\[2mm]
&= \frac{1}{2} A''' \left[I_\parallel + I_\perp - (I_\parallel - I_\perp) \cos 2\omega_L(t - t_0) \right] e^{-\frac{t-t_0}{\tau}} \\[2mm]
&= \frac{1}{2} A''' (I_\parallel + I_\perp) \left[1 - \frac{I_\parallel - I_\perp}{I_\parallel + I_\perp} \cos 2\omega_L(t - t_0) \right] e^{-\frac{t-t_0}{\tau}} , \\[2mm]
&= A' \left[1 - P \cos 2\omega_L(t - t_0) \right] e^{-\frac{t-t_0}{\tau}} , \qquad (3.217)
\end{aligned}
$$

where A' is a new proportionality constant and P is the degree of polarization defined by equation Z-3.8.3.

B.

SUPPOSE THAT EITHER THE INCIDENT LIGHT IS PLANE POLARIZED ALONG **H** (THE Z AXIS) OR THE RESONANCE FLUORESCENCE IS ANALYZED FOR POLARIZATION ALONG **H** (THE Z AXIS). DISCUSS HOW THE RESONANCE FLUORESCENCE SIGNAL DEPENDS ON THE MAGNETIC FIELD STRENGTH FOR THESE EXCITATION-DETECTION GEOMETRIES.

We consider two arrangements of the polarization vectors for excitation and detection: (*i*) the incident photon is polarized parallel to **H** (along the Z axis) and the polarization direction of the detected photon is arbitary (along the X or Z axes), or (*ii*) the incident polarization is arbitrary (along the Y or Z axes) and the detected photon is polarized parallel to **H** (along the Z axis). As before, the incident photon propagates in the X direction and the detected photon propagates in the Y direction.

For these two schemes, the resonance fluorescence signal at time t is given by

$$(i) \qquad I(H, t - t_0) = A \overline{\Phi^2_{Zg} \Phi^2_{F'h}}(H, t - t_0) \, e^{-\frac{t-t_0}{\tau}}, \tag{3.218}$$

where $F' = Z$ or X, and

$$(ii) \qquad I(H, t - t_0) = A \overline{\Phi^2_{Fg} \Phi^2_{Zh}}(H, t - t_0) \, e^{-\frac{t-t_0}{\tau}}, \tag{3.219}$$

where $F = Z$ or Y.

From equations 3.218 and 3.219, we see that there are three possible combinations of excitation and detection geometries, (Z, Z), (Z, X), and (Y, Z). We will evaluate $I(H, t - t_0)$ for each of them.

(a) (Z, Z)

$$
\begin{aligned}
I(H, t - t_0) &= A \overline{\Phi^2_{Zg} \Phi^2_{Zh}}(H, t - t_0) \, e^{-\frac{t-t_0}{\tau}} \\[2mm]
&= A \overline{\Phi^2_{Z(t_0)g} \Phi^2_{Z(t)h}} \, e^{-\frac{t-t_0}{\tau}} \\[2mm]
&= A \overline{\Phi^2_{Z(t_0)g} \Phi^2_{Z(t_0)h}} \, e^{-\frac{t-t_0}{\tau}} \\[2mm]
&= A' I_{\parallel} \, e^{-\frac{t-t_0}{\tau}},
\end{aligned}
\tag{3.220}
$$

where we used equation 3.210 for $\Phi^2_{Z(t)h}$ and equation Z-3.8.1 for $I_{\parallel}$.

(b) (Z, X)

$$
\begin{aligned}
I(H, t - t_0) &= A \overline{\Phi^2_{Zg} \Phi^2_{Xh}}(H, t - t_0) \, e^{-\frac{t-t_0}{\tau}} \\[2mm]
&= A \overline{\Phi^2_{Z(t_0)g} \Phi^2_{X(t)h}} \, e^{-\frac{t-t_0}{\tau}} \\[2mm]
&= A \overline{\Phi^2_{Z(t_0)g} \left[\Phi_{X(t_0)h} \cos \omega_L (t - t_0) + \Phi_{Y(t_0)h} \sin \omega_L (t - t_0) \right]^2} \, e^{-\frac{t-t_0}{\tau}} \\[2mm]
&= A \left[\overline{\Phi^2_{Z(t_0)g} \Phi^2_{X(t_0)h}} \cos^2 \omega_L (t - t_0) \; + \; \overline{\Phi^2_{Z(t_0)g} \Phi^2_{Y(t_0)h}} \sin^2 \omega_L (t - t_0) \right. \\[2mm]
&\qquad \left. + \, 2 \overline{\Phi^2_{Z(t_0)g} \Phi_{X(t_0)h} \Phi_{Y(t_0)h}} \sin \omega_L (t - t_0) \cos \omega_L (t - t_0) \right] \, e^{-\frac{t-t_0}{\tau}}
\end{aligned}
$$

$$= A \left[I_\perp \cos^2 \omega_L(t - t_0) + I_\perp \sin^2 \omega_L(t - t_0) \right] e^{-\frac{t-t_0}{\tau}}$$

$$= B I_\perp e^{-\frac{t-t_0}{\tau}}, \tag{3.221}$$

where we used equations 3.211, Z-3.8.1, and Z-3.8.11.

(c) (Y, Z)

$$I(H, t - t_0) = A \overline{\Phi_{Yg}^2 \Phi_{Zh}^2}(H, t - t_0) \, e^{-\frac{t-t_0}{\tau}}$$

$$= A \overline{\Phi_{Y(t_0)g}^2 \, \Phi_{Z(t)h}^2} \, e^{-\frac{t-t_0}{\tau}}$$

$$= A \overline{\Phi_{Y(t_0)g}^2 \, \Phi_{Z(t_0)h}^2} \, e^{-\frac{t-t_0}{\tau}}$$

$$= C I_\perp \, e^{-\frac{t-t_0}{\tau}}, \tag{3.222}$$

where we used equations 3.210 and Z-3.8.1.

In all three geometries, we see that the fluorescence signal does not depend on the magnetic field strength. The physical reason for this lack of field dependence is that there is no average torque acting on the oscillators.

C.

INSTEAD OF A LIGHT PULSE OF SHORT DURATION, CONSIDER A CONTINUOUS LIGHT BEAM. SHOW THAT $I(H)$ IS A LORENZTIAN FUNCTION OF THE MAGNETIC FIELD STRENGTH WITH A HALF-WIDTH AT HALF-MAXIMUM OF

$$H_{\frac{1}{2}} = \frac{\hbar}{2\mu_0 g \tau}.$$

From part A, we know how the intensity I varies with the field $\mathbf{H}$ for a short light pulse. To extend the treatment to very long pulses ($t \gg \tau$), we need to integrate equation 3.217 from $-\infty$ to t:

$$I(H) = \int_{-\infty}^{t} I(H, t - t_0) \mathrm{d}t_0$$

$$= A' \int_{-\infty}^{t} [1 - P \cos 2\omega_L(t - t_0)] \, e^{-\frac{t-t_0}{\tau}} \mathrm{d}t_0$$

$$= A' \left[\tau - P \int_{-\infty}^{t} \cos 2\omega_L(t - t_0) \, e^{-\frac{t-t_0}{\tau}} \mathrm{d}t_0 \right]$$

$$= A' \left[\tau - P \int_0^\infty \cos 2\omega_L x \, e^{-\frac{x}{\tau}} \mathrm{d}x \right]$$

$$= A'\tau \left[1 - \frac{P}{1 + 4\omega_L^2 \tau^2} \right]$$

$$= A'\tau \left[1 - \frac{P}{1 + \frac{4\mu_0^2 g^2 \tau^2}{\hbar^2} H^2} \right], \tag{3.223}$$

where we have set $t - t_0 = x$ and used equation Z-3.9.1 and the integral

$$\int_0^\infty e^{-px} \cos qx \, \mathrm{d}x = \frac{p}{p^2 + q^2} \, .$$

The second term in equation 3.223 is a Lorentzian function of H, with the full width at half-maximum, $H_{\frac{1}{2}}$, given by

$$\frac{1}{2} = \frac{1}{1 + \frac{4\mu_0^2 g^2 \tau^2}{\hbar^2} H_{\frac{1}{2}}^2}. \tag{3.224}$$

Solving this equation for $H_{\frac{1}{2}}$, we obtain

$$H_{\frac{1}{2}} = \frac{\hbar}{2\mu_0 g \tau}. \tag{3.225}$$

D.

SUPPOSE THAT THE INTENSITY OF THE INCIDENT LIGHT BEAM IS MODULATED AT THE FREQUENCY ω ACCORDING TO

$$I(t_0) = I_0 \left(\frac{1}{2} + \frac{1}{2} \cos \omega t_0 \right). \tag{3.226}$$

DETERMINE THE STEADY-STATE RESPONSE OF THE SYSTEM. CONSIDER WHAT HAPPENS TO THE RESONANCE FLUORESCENCE IF THE MODULATION FREQUENCY OR THE MAGNETIC FIELD STRENGTH IS SWEPT. DESIGN AN EXPERIMENT TO DETERMINE g AND τ INDEPENDENTLY.

The steady-state response of the system can be found by convoluting the intensity of the incident light beam (equation 3.226) with the intensity of the resonance fluorescence for a short pulse (equation 3.217) from $t_0 = -\infty$ to $t_0 = t$,

$$I(H, \omega, t) = \int_{-\infty}^t I_0 \left(\frac{1}{2} + \frac{1}{2} \cos \omega t_0 \right) A' \left[1 - P \cos 2\omega_L (t - t_0) \right] e^{-\frac{t-t_0}{\tau}} \mathrm{d}t_0$$

$$= \frac{I_0 A'}{2} \left[\int_{-\infty}^t e^{-\frac{t-t_0}{\tau}} \mathrm{d}t_0 - P \int_{-\infty}^t \cos 2\omega_L (t - t_0) e^{-\frac{t-t_0}{\tau}} \mathrm{d}t_0 \right.$$

$$+ \int_{-\infty}^{t} \cos \omega t_0 \; e^{-\frac{t-t_0}{\tau}} \mathrm{d}t_0 - P \int_{-\infty}^{t} \cos \omega_L t_0 \cos 2\omega_L (t - t_0) \; e^{-\frac{t-t_0}{\tau}} \mathrm{d}t_0 \bigg]$$

$$= \frac{I_0 A'}{2} \left[\tau \left(1 - \frac{P}{1 + 4\omega_L^2 \tau^2} \right) + \int_{-\infty}^{t} \cos \omega t_0 \; e^{-\frac{t-t_0}{\tau}} \mathrm{d}t_0 \right.$$

$$\left. -P \int_{-\infty}^{t} \cos \omega t_0 \cos 2\omega_L (t - t_0) \; e^{-\frac{t-t_0}{\tau}} \mathrm{d}t_0 \right], \tag{3.227}$$

where we use the result of part C to evaluate the first two integrals. The third integral has the value

$$\int_{-\infty}^{t} \cos \omega t_0 \; e^{-\frac{t-t_0}{\tau}} \mathrm{d}t_0 \;\; = \;\; \int_{0}^{\infty} \cos \omega (t - x) \; e^{-\frac{x}{\tau}} \mathrm{d}x$$

$$= \;\; \frac{1}{\tau^{-2} + \omega^2} \left(\frac{1}{\tau} \cos \omega t + \omega \sin \omega t \right), \tag{3.228}$$

where we have set $t - t_0 = x$ and used the relation

$$\int_{0}^{\infty} e^{-px} \cos(qx + \lambda) \mathrm{d}x = \frac{1}{p^2 + q^2} \left(p \cos \lambda - q \sin \lambda \right).$$

The fourth integral in equation 3.227 has the value

$$\int_{-\infty}^{t} \cos \omega t_0 \cos 2\omega_L (t - t_0) \; e^{-\frac{t-t_0}{\tau}} \mathrm{d}t_0$$

$$= \;\; \int_{-\infty}^{t} \cos \omega (t - x) \cos 2\omega_L x \; e^{-\frac{x}{\tau}} \mathrm{d}x$$

$$= \;\; \frac{1}{2} \int_{0}^{\infty} \left[\cos \left(2\omega_L x - \omega x + \omega t \right) + \cos \left(2\omega_L x + \omega x - \omega t \right) \right] \; e^{-\frac{x}{\tau}} \mathrm{d}x$$

$$= \;\; \frac{1}{2} \left[\frac{\tau^{-1} \cos \omega t - (2\omega_L - \omega) \sin \omega t}{\tau^{-2} + (2\omega_L - \omega)^2} + \frac{\tau^{-1} \cos \omega t + (2\omega_L + \omega) \sin \omega t}{\tau^{-2} + (2\omega_L + \omega)^2} \right]$$

$$= \;\; \frac{\tau}{2} \left[\frac{\cos \omega t - (2\omega_L - \omega)\tau \sin \omega t}{1 + (2\omega_L - \omega)^2 \tau^2} + \frac{\cos \omega t + (2\omega_L + \omega)\tau \sin \omega t}{1 + (2\omega_L + \omega)^2 \tau^2} \right], \tag{3.229}$$

where we have used the identity

$$\cos \alpha \cos \beta = \frac{1}{2} \left[\cos(\alpha + \beta) + \cos(\alpha - \beta) \right].$$

Inserting equations 3.228 and 3.229 into equation 3.227, we obtain

$$
\begin{aligned}
I(H,\omega,t) \;=\;& \frac{I_0 A'\tau}{2}\left\{ 1 - \frac{P}{1+4\omega_L^2\tau^2} + \frac{\cos\omega t + \omega\tau\sin\omega t}{1+\omega^2\tau^2} \right.\\[2ex]
& \left. - \frac{P}{2}\left[\frac{\cos\omega t - (2\omega_L-\omega)\tau\sin\omega t}{1+(2\omega_L-\omega)^2\tau^2} + \frac{\cos\omega t + (2\omega_L+\omega)\tau\sin\omega t}{1+(2\omega_L+\omega)^2\tau^2} \right] \right\}\\[3ex]
\;=\;& \frac{I_0 A'\tau}{2}\left\{ \left[1 - \frac{P}{1+4\omega_L^2\tau^2} \right] \right.\\[3ex]
& + \left[\frac{1}{1+\omega^2\tau^2} - \frac{P}{2[1+(2\omega_L-\omega)^2\tau^2]} - \frac{P}{2[1+(2\omega_L+\omega)^2\tau^2]} \right]\cos\omega t\\[3ex]
& \left. + \left[\frac{\omega\tau}{1+\omega^2\tau^2} + \frac{P(2\omega_L-\omega)\tau}{2[1+(2\omega_L-\omega)^2\tau^2]} - \frac{P(2\omega_L+\omega)\tau}{2[1+(2\omega_L+\omega)^2\tau^2]} \right]\sin\omega t \right\}.
\end{aligned}
\tag{3.230}
$$

The result in equation 3.230 consists of the time-independent Lorentzian obtained in part C and an oscillatory component with a frequency ω. If ω is swept, three resonances occur at frequencies

$$
\begin{aligned}
\omega \;&=\; 0,\quad \pm 2\omega_L\\[2ex]
\;&=\; 0,\quad \pm 2\frac{\mu_0 g H}{\hbar}\;.
\end{aligned}
\tag{3.231}
$$

If H is swept instead, the resonances occur at

$$
\omega_L = 0,\quad \pm\frac{1}{2}\omega
\tag{3.232}
$$

or

$$
H = 0,\quad \pm\frac{1}{2}\frac{\hbar\omega}{\mu_0 g}\;.
\tag{3.233}
$$

To determine g and τ, we may use either pulsed or continuous light. Using pulsed light, we can determine τ from the decay of the fluorescence, as described in part B. We can then perform a Zeeman quantum beat experiment, as described in part A, to determine ω_L (and hence g) from a fit of the oscillatory signal. With high-quality data, it is possible to obtain both g and τ from a fit of the quantum beats.

If we use a continuous incident light beam, we can determine $H_{\frac{1}{2}}$ from the unmodulated fluorescence (i.e., from the Hanle effect) and thereby obtain the value of $g\tau$. Next, we measure the modulated fluorescence, as described in part D. From equation 3.230, we know that

$$
I(H=0,\omega,t) \;=\; \frac{I_0 A'\tau}{2}\left[1 - P + \frac{(1-P)\cos\omega t}{1+\omega^2\tau^2} + \frac{\omega\tau(1-P)\sin\omega t}{1+\omega^2\tau^2} \right]
$$

$$= \frac{I_0 A' \tau (1 - P)}{2} \left[1 + \frac{\cos \omega t + \omega \tau \sin \omega t}{1 + \omega^2 \tau^2} \right]. \tag{3.234}$$

The ratio of the oscillatory component ($\cos \omega t$ or $\omega \tau \sin \omega t$) for $H = 0$ to the static component is given by

$\left(1 + \omega^2 \tau^2 \right)^{-1}$, from which τ can be extracted. Knowing both $g\tau$ and τ allows us to determine g.

ection[Application 10   Correlation Functions in Molecular Spectroscopy]APPLICATION 10
CORRELATION FUNCTIONS IN MOLECULAR SPECTROSCOPY

IN THE TIME-INDEPENDENT PICTURE, THE LINESHAPE FUNCTION IS GIVEN BY

$$I(\omega) = 3 \sum_i \sum_f \rho_i \left| \langle f | \hat{\mathbf{e}} \cdot \boldsymbol{\mu} | i \rangle \right|^2 \delta(\omega_{fi} - \omega), \tag{3.235}$$

WHERE THE INITIAL STATE $|i\rangle$ HAS ENERGY E_i, THE FINAL STATE $|f\rangle$ HAS ENERGY E_f, ρ_i IS THE PROB-
ABILITY OF FINDING THE SYSTEM IN STATE $|i\rangle$, AND $\langle f | \hat{\mathbf{e}} \cdot \boldsymbol{\mu} | i \rangle$ IS THE TRANSITION DIPOLE MOMENT
MATRIX ELEMENT. IN THE HEISENBERG PICTURE, THE TIME-DEPENDENT TRANSITION DIPOLE MOMENT
OPERATOR CAN BE WRITTEN AS

$$\boldsymbol{\mu}(t) = e^{\frac{i\mathcal{H}t}{\hbar}} \boldsymbol{\mu}(0) e^{\frac{-i\mathcal{H}t}{\hbar}}. \tag{3.236}$$

A.

SHOW THAT THE LINESHAPE FACTOR MAY BE WRITTEN AS

$$I(\omega) = \frac{3}{2\pi} \int_{-\infty}^{\infty} e^{-i\omega t} \sum_i \rho_i \langle i | \hat{\mathbf{e}} \cdot \boldsymbol{\mu}(0) \hat{\mathbf{e}} \cdot \boldsymbol{\mu}(t) | i \rangle \, \mathrm{d}t \tag{3.237}$$

We start by replacing the delta function in equation 3.235 by its Fourier expansion,

$$\delta(\omega_{fi} - \omega) = \frac{1}{2\pi} \int_{-\infty}^{\infty} e^{i(\omega_{fi} - \omega)t} \mathrm{d}t. \tag{3.238}$$

Introducing the identity Z-3.10.5 gives

$$I(\omega) = \frac{3}{2\pi} \sum_i \sum_f \rho_i \langle i | \hat{\mathbf{e}} \cdot \boldsymbol{\mu}(0) | f \rangle \langle f | \hat{\mathbf{e}} \cdot \boldsymbol{\mu}(0) | i \rangle \int_{-\infty}^{\infty} e^{i(\omega_{fi} - \omega)t} \mathrm{d}t. \tag{3.239}$$

Factoring the exponential and introducing the Bohr condition (Z-3.10.1), we obtain

$$I(\omega) = \frac{3}{2\pi} \int_{-\infty}^{\infty} e^{-i\omega t} \sum_i \sum_f \rho_i \left\langle i \left| \hat{\mathbf{e}} \cdot \boldsymbol{\mu}(0) \right| f \right\rangle \left\langle f \left| \hat{\mathbf{e}} \cdot \boldsymbol{\mu}(0) \, e^{i(E_f - E_i)t/\hbar} \right| i \right\rangle \, \mathrm{d}t. \tag{3.240}$$

Introducing equation Z-3.10.4, we obtain

$$I(\omega) = \frac{3}{2\pi} \int_{-\infty}^{\infty} e^{-i\omega t} \sum_i \sum_f \rho_i \left\langle i \left| \hat{\mathbf{e}} \cdot \boldsymbol{\mu}(0) \right| f \right\rangle \left\langle f \left| e^{i\hat{\mathcal{H}}t/\hbar} \, \hat{\mathbf{e}} \cdot \boldsymbol{\mu}(0) \, e^{-i\hat{\mathcal{H}}t/\hbar} \right| i \right\rangle \, \mathrm{d}t$$

$$= \frac{3}{2\pi} \int_{-\infty}^{\infty} e^{-i\omega t} \sum_i \sum_f \rho_i \langle i | \hat{\mathbf{e}} \cdot \boldsymbol{\mu}(0) | f \rangle \langle f | \hat{\mathbf{e}} \cdot \boldsymbol{\mu}(t) | i \rangle \, \mathrm{d}t. \tag{3.241}$$

Recognizing the identity operator in the last equation,

$$\sum_f |f\rangle \langle f| = 1,$$

gives the result of equation 3.237.

B.

BECAUSE $I(\omega)$ IS A REAL QUANTITY, THAT IS, $I(\omega) = I^*(\omega)$, SHOW THAT

$$G(t) = G^*(-t).$$

The dipole correlation function $G(t)$ is defined by

$$G(t) = \langle \boldsymbol{\mu}(0) \cdot \boldsymbol{\mu}(t) \rangle, \tag{3.242}$$

where the brackets indicate an ensemble average. Assuming that the system is isotropic and taking the inverse Fourier transform of equation 3.237, we can write $G(t)$ as

$$G(t) = \int_{band} e^{i\omega t} I(\omega) d\omega. \tag{3.243}$$

The complex conjugate of $G(t)$ is

$$
\begin{aligned}
G^*(t) &= \int_{band} e^{-i\omega t} I^*(\omega) d\omega \\[2mm]
&= \int_{band} e^{-i\omega t} I(\omega) d\omega \\[2mm]
&= G(-t),
\end{aligned}
\tag{3.244}
$$

It follows that

$$\mathcal{R}e[G(t)] = \mathcal{R}e[G(-t)] \tag{3.245}$$

and

$$\mathcal{I}m[G(t)] = -\mathcal{I}m[G(-t)]. \tag{3.246}$$

C.

SUPPOSE THAT THE DIPOLE CORRELATION FUNCTION VARIES SINUSOIDALLY IN TIME WITH A FREQUENCY ω_0, THAT IS, $G(t) = e^{i\omega_0 t}$. FIND $I(\omega)$ AND DISCUSS ITS FORM.

From equations Z-3.10.8 and Z-3.10.9 we know that

$$I(\omega) = \frac{1}{2\pi} \int_{-\infty}^{\infty} e^{-i\omega t} G(t) dt. \tag{3.247}$$

If we introduce the assumed form for $G(t)$ into equation 3.247, we obtain

$$I(\omega) = \frac{1}{2\pi} \int_{-\infty}^{\infty} e^{-i\omega t} e^{i\omega_0 t} dt$$

$$= \frac{1}{2\pi} \int_{-\infty}^{\infty} e^{i(\omega_0 - \omega)t} dt$$

$$= \delta(\omega_0 - \omega). \tag{3.248}$$

The lineshape factor is a delta function that peaks at $\omega = \omega_0$.

D.

SUPPOSE THAT $G(t)$ HAS A SINUSOIDAL VARIATION IN TIME WITH FREQUENCY ω_0 BUT ALSO DECREASES EXPONENTIALLY WITH A HALF-LIFE τ, THAT IS,

$$G(t) = \tau^{-1} e^{\frac{-|t|}{\tau}} e^{i\omega_0 t}.$$

FIND $I(\omega)$ AND DISCUSS ITS FORM.

Substituting $G(t)$ into equation Z-3.10.18, we obtain

$$I(\omega) = \frac{1}{\pi} \mathcal{R}e \left[\int_0^{\infty} e^{-i\omega t} \tau^{-1} e^{\frac{-|t|}{\tau}} e^{i\omega_0 t} dt \right]$$

$$= \frac{1}{\pi} \mathcal{R}e \left[\int_0^{\infty} \tau^{-1} e^{(i(\omega_0 - \omega) - \tau^{-1})t} dt \right]$$

$$= \frac{1}{\tau\pi} \mathcal{R}e \left[\frac{e^{(i(\omega_0 - \omega) - \tau^{-1})t}}{i(\omega_0 - \omega) - \tau^{-1}} \Big|_0^{\infty} \right]$$

$$= \frac{1}{\tau\pi} \mathcal{R}e \left[\frac{1}{\tau^{-1} - i(\omega_0 - \omega)} \right]$$

$$= \frac{1}{\pi} \frac{1}{1 + (\omega_0 - \omega)^2 \tau^2}$$

$$= \frac{1}{\pi\tau^2} \frac{1}{(\omega - \omega_0)^2 + \frac{1}{\tau^2}}. \tag{3.249}$$

$I(\omega)$ is a Lorentzian function centered at ω_0 with a full width at half-maximum of $2/\tau$.

E.

AS ANOTHER EXAMPLE, CONSIDER THE CLASSICAL INFRARED SPECTRUM OF AN EQUILIBRIUM DISTRIBUTION OF LINEAR MOLECULES WITH A MOMENT OF INERTIA I AND TEMPERATURE T. FIRST SHOW THAT

THE PROBABILITY OF FINDING A LINEAR MOLECULE ROTATING WITH AN ANGULAR FREQUENCY BETWEEN Ω AND $\Omega + d\Omega$ IS GIVEN BY

$$f(\Omega)d\Omega = \left(\frac{I}{kT}\right)\Omega\; e^{-\frac{I\Omega^2}{2kT}}d\Omega, \tag{3.250}$$

FOR $0 \leq \Omega \leq \infty$.

Treating the rotational quantum number as a continuous variable, the distribution in J is given by

$$f(J)dJ = \frac{(2J+1)h^2}{8\pi^2 IkT}\; e^{\frac{-J(J+1)h^2}{8\pi^2 IkT}}dJ. \tag{3.251}$$

Here the transition dipole moment vector lies along the molecular axis, and if the latter rotates at an angular frequency Ω, then

$$\hat{\boldsymbol{\mu}}(0) \cdot \hat{\boldsymbol{\mu}}(t) = \cos\Omega t. \tag{3.252}$$

By identifying the rotational energy of a linear rotor with the classical rotational energy, $\frac{1}{2}I\Omega^2$, we obtain

$$\frac{J(J+1)h^2}{8\pi^2 I} = \frac{1}{2}\; I\; \Omega^2 \tag{3.253}$$

and

$$(2J+1)h^2\; dJ = 8\pi^2\; I^2\; \Omega\; d\Omega \tag{3.254}$$

for $\Omega \geq 0$. Substituting equations 3.251 and 3.254 into the relation

$$f(J)dJ = f(\Omega)d\Omega, \tag{3.255}$$

we obtain

$$\begin{aligned}
f(\Omega)d\Omega &= \frac{(2J+1)h^2}{8\pi^2 IkT}\; e^{\frac{-J(J+1)h^2}{8\pi^2 IkT}}dJ \\[2em]
&= \frac{8\pi^2\; I^2\; \Omega}{8\pi^2\; I\; kT}\; e^{\frac{-I\Omega^2}{2kT}}\; d\Omega \\[2em]
&= \frac{I\; \Omega}{kT}\; e^{\frac{-I\Omega^2}{2kT}}\; d\Omega
\end{aligned} \tag{3.256}$$

for $0 \leq \Omega \leq \infty$.

NEXT, SHOW THAT

$$I(\omega) = \left(\frac{I}{kT}\right)|\omega|\; e^{\frac{-I\omega^2}{2kT}}.$$

Hint: DO NOT EXPLICITLY EVALUATE THE EQUATION; INSTEAD, THINK FOURIER TRANSFORM PAIRS.

The normalized dipole correlation function is found by taking the ensemble average over equation 3.252:

$$\langle \hat{\boldsymbol{\mu}}(0) \cdot \hat{\boldsymbol{\mu}}(t) \rangle = \mathcal{R}e \int_0^\infty \hat{\boldsymbol{\mu}}(0) \cdot \hat{\boldsymbol{\mu}}(t) f(\Omega) \mathrm{d}(\Omega)$$

$$= \left(\frac{I}{kT} \right) \mathcal{R}e \left[\int_0^\infty \Omega \, e^{-\frac{I\Omega^2}{2kT}} \, e^{i\Omega t} \mathrm{d}\Omega \right]. \tag{3.257}$$

Here we have written $\langle \hat{\boldsymbol{\mu}}(0) \cdot \hat{\boldsymbol{\mu}}(t) \rangle$ as the real part of the complex correlation function. The latter is given by

$$\langle \hat{\boldsymbol{\mu}}(0) \cdot \hat{\boldsymbol{\mu}}(t) \rangle = \int_{band} e^{i\omega t} I(\omega) \mathrm{d}\omega. \tag{3.258}$$

In equation 3.257, Ω is a dummy variable that can be changed to ω, so that

$$\langle \hat{\boldsymbol{\mu}}(0) \cdot \hat{\boldsymbol{\mu}}(t) \rangle = \left(\frac{I}{kT} \right) \mathcal{R}e \left[\int_0^\infty \omega \, e^{-\frac{I\omega^2}{2kT}} \, e^{i\omega t} \mathrm{d}\omega \right]. \tag{3.259}$$

Comparing equation 3.258 with 3.259, we identify

$$I(\omega) = \left(\frac{I}{kT} \right) \omega \, e^{-\frac{I\omega^2}{2kT}}. \tag{3.260}$$

Considering that, classically, the angular frequency can be either positive or negative, we can write

$$I(\omega) = \left(\frac{I}{kT} \right) |\omega| \, e^{-\frac{I\omega^2}{2kT}}. \tag{3.261}$$

Chapter 4

COUPLING OF MORE THAN TWO ANGULAR MOMENTUM VECTORS

PROBLEM SET 2

1.   THE VECTOR MODEL ALLOWS A SIMPLE GEOMETRIC INTERPRETATION OF THE EXPRESSION

$$\langle j_{12} j_3 j | j_1 j_{23} j \rangle^2.$$

THE PROBABILITY THAT A SYSTEM PREPARED IN A STATE OF THE COUPLING SCHEME $\mathbf{j} = \mathbf{j}_{12} + \mathbf{j}_3$ WILL BE FOUND TO BE IN A STATE OF THE COUPLING SCHEME $\mathbf{j} = \mathbf{j}_1 + \mathbf{j}_{23}$ IS

$$P = \langle j_{12} j_3 j | j_1 j_{23} j \rangle^2$$

$$= (2j_{12} + 1)(2j_{23} + 1) \left\{ \begin{array}{ccc} j_1 & j_2 & j_{12} \\ j_3 & j & j_{23} \end{array} \right\}^2. \tag{4.1}$$

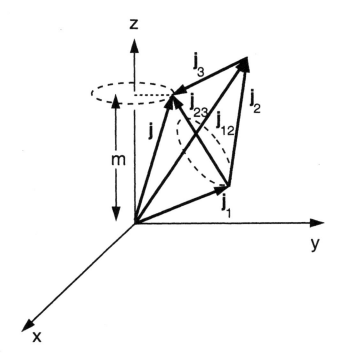

Figure 4.1: Vector model for the coupling of three angular momenta.

ACCORDING TO FIGURE 4.1, $\langle j_{12} j_3 j | j_1 j_{23} j \rangle$ IS INDEPENDENT OF THE VALUE OF m. THE SIX ANGULAR MOMENTA $\mathbf{j}_1$, $\mathbf{j}_2$, $\mathbf{j}_3$, $\mathbf{j}_{12}$, $\mathbf{j}_{23}$, AND $\mathbf{j}$ MAY BE REGARDED AS FORMING THE SIDES OF AN IRREGULAR TETRAHEDRON WHOSE VOLUME V IS GIVEN BY

$$V = \frac{1}{3} \left[\frac{1}{2} (\mathbf{j}_{12} \times \mathbf{j}) \cdot \mathbf{j}_1 \right]. \tag{4.2}$$

IF THE MAGNITUDES OF $\mathbf{j}_1$, $\mathbf{j}_2$, $\mathbf{j}_3$, $\mathbf{j}_{12}$, AND $\mathbf{j}$ ARE FIXED, THE MAGNITUDE OF $\mathbf{j}_{23}$ IS FREE TO VARY IN SUCH A WAY THAT THE LOCUS OF $\mathbf{j}_{23}$ DESCRIBES A CIRCLE IN SPACE CENTERED ABOUT $\mathbf{j}_{12}$. THEN THE DIHEDRAL ANGLE ϕ BETWEEN THE TWO PLANES CONTAINING $(\mathbf{j}, \mathbf{j}_{12}, \mathbf{j}_3)$ AND $(\mathbf{j}_1, \mathbf{j}_2, \mathbf{j}_{12})$ VARIES

BETWEEN 0 AND 2π AT A UNIFORM RATE. THE UNIT VECTORS NORMAL TO THESE TWO PLANES ARE $(\mathbf{j}_{12} \times \mathbf{j})/|\mathbf{j}_{12} \times \mathbf{j}|$ AND $(\mathbf{j}_{12} \times \mathbf{j}_1)/|\mathbf{j}_{12} \times \mathbf{j}_1|$, RESPECTIVELY. HENCE

$$\cos\phi = \frac{(\mathbf{j}_{12} \times \mathbf{j}) \cdot (\mathbf{j}_{12} \times \mathbf{j}_1)}{|\mathbf{j}_{12} \times \mathbf{j}||\mathbf{j}_{12} \times \mathbf{j}_1|} \tag{4.3}$$

AND

$$|\sin\phi| = \frac{|(\mathbf{j}_{12} \times \mathbf{j}) \times (\mathbf{j}_{12} \times \mathbf{j}_1)|}{|\mathbf{j}_{12} \times \mathbf{j}||\mathbf{j}_{12} \times \mathbf{j}_1|}. \tag{4.4}$$

1A. DIFFERENTIATE EQUATION 4.3 WITH RESPECT TO TIME TO SHOW THAT

$$\frac{d|\mathbf{j}_{23}|}{dt} = \frac{(\mathbf{j}_{12} \times \mathbf{j}) \cdot \mathbf{j}_1}{|\mathbf{j}_{12}||\mathbf{j}_{23}|}\frac{d\phi}{dt}, \tag{4.5}$$

WHERE $d\phi/dt$ MAY BE REPLACED BY 2π. *Hint:* MAKE USE OF THE LAW OF COSINES

$$|\mathbf{j}_{23}|^2 = |\mathbf{j}_1|^2 + |\mathbf{j}|^2 - 2\mathbf{j}_1 \cdot \mathbf{j} \tag{4.6}$$

AND

$$|\mathbf{j}_1|^2 = |\mathbf{j}_{23}|^2 + |\mathbf{j}|^2 - 2\mathbf{j}_{23} \cdot \mathbf{j} \tag{4.7}$$

TO MAKE THE IDENTIFICATION

$$\frac{d(\mathbf{j}_1 \cdot \mathbf{j})}{dt} = \frac{-d(\mathbf{j}_{23} \cdot \mathbf{j})}{dt} = -|\mathbf{j}_{23}|\frac{d|\mathbf{j}_{23}|}{dt}. \tag{4.8}$$

Using vector algebra, we can rewrite equations 4.3 and 4.4 as

$$\cos\phi = \frac{1}{|\mathbf{j}_{12} \times \mathbf{j}||\mathbf{j}_{12} \times \mathbf{j}_1|}\{|\mathbf{j}_{12}|^2\mathbf{j} \cdot \mathbf{j}_1 - (\mathbf{j}_{12} \cdot \mathbf{j}_1)(\mathbf{j}_{12} \cdot \mathbf{j})\} \tag{4.9}$$

and

$$|\sin\phi| = \frac{1}{|\mathbf{j}_{12} \times \mathbf{j}||\mathbf{j}_{12} \times \mathbf{j}_1|}|\mathbf{j}_{12}|\left[(\mathbf{j}_{12} \times \mathbf{j}) \cdot \mathbf{j}_1\right]. \tag{4.10}$$

By differentiating equation 4.9 with respect to time and noting that $|\mathbf{j}_{12} \times \mathbf{j}|$, $|\mathbf{j}_{12} \times \mathbf{j}_1|$, $|\mathbf{j}_{12}|$, $\mathbf{j}_{12} \cdot \mathbf{j}$, and $\mathbf{j}_{12} \cdot \mathbf{j}_1$ do not vary with time, we obtain

$$-\sin\phi\frac{d\phi}{dt} = \frac{1}{|\mathbf{j}_{12} \times \mathbf{j}||\mathbf{j}_{12} \times \mathbf{j}_1|}|\mathbf{j}_{12}|^2\frac{d(\mathbf{j} \cdot \mathbf{j}_1)}{dt}. \tag{4.11}$$

Inserting equation 4.10 into equation 4.11, we get

$$\frac{d(\mathbf{j} \cdot \mathbf{j}_1)}{dt} = -\frac{1}{|\mathbf{j}_{12}|}\left[(\mathbf{j}_{12} \times \mathbf{j}) \cdot \mathbf{j}_1\right]\frac{d\phi}{dt}. \cdot \tag{4.12}$$

From equation 4.6 we obtain

$$-\frac{d(\mathbf{j}_1 \cdot \mathbf{j})}{dt} = \frac{1}{2}\frac{d|\mathbf{j}_{23}|^2}{dt}$$

$$= |\mathbf{j}_{23}|\frac{d|\mathbf{j}_{23}|}{dt}, \tag{4.13}$$

because $|\mathbf{j}_1|^2$ and $|\mathbf{j}|^2$ do not vary with time. Finally, inserting equation 4.13 into equation 4.12, we obtain the desired result.

1B.   THE PROBABILITY DENSITY FOR A GIVEN VALUE OF $|\mathbf{j}_{23}|$ MAY THEN BE FOUND FROM

$$P(j_{23}) = 2 \left(\frac{d|\mathbf{j}_{23}|}{dt} \right)^{-1}, \tag{4.14}$$

WHERE THE FACTOR OF 2 ARISES FROM THE FACT THAT $\mathbf{j}_{23}$ TAKES ON ANY OF ITS VALUES TWICE AS ϕ GOES FROM 0 TO 2π. HENCE SHOW THAT IN THE CLASSICAL LIMIT FOR LARGE ANGULAR MOMENTA, THE SQUARE OF THE 6-j SYMBOL HAS THE LIMITING VALUE

$$\begin{Bmatrix} j_1 & j_2 & j_{12} \\ j_3 & j & j_{23} \end{Bmatrix}^2 = \frac{1}{4\pi(\mathbf{j}_{12} \times \mathbf{j}) \cdot \mathbf{j}_1} = \frac{1}{24\pi V}. \tag{4.15}$$

Substituting equation 4.5 into equation 4.14 and equating $d\phi/dt$ with 2π, we get the semiclassical value for $P(j_{23})$,

$$P(j_{23}) = \pi^{-1}|\mathbf{j}_{12}||\mathbf{j}_{23}| \frac{1}{(\mathbf{j}_{12} \times \mathbf{j}) \cdot \mathbf{j}_1}. \tag{4.16}$$

The semiclassical (vector-model) values for $|\mathbf{j}_{12}|$ and $|\mathbf{j}_{23}|$ are given by

$$|\mathbf{j}_{12}| = j_{12} + \frac{1}{2} \quad \text{and} \quad |\mathbf{j}_{23}| = j_{23} + \frac{1}{2}. \tag{4.17}$$

Substituting equations 4.2, 4.16, and 4.17 into equation 4.1, we get the desired semiclassical expression for the square of the 6-j symbol.

2.   PROVE BY DIAGRAMMATIC MEANS THAT

$$\sum_{m_4 m_5 m_6} (-1)^{j_4 + m_4 + j_5 + m_5 + j_6 + m_6} \begin{pmatrix} j_1 & j_5 & j_6 \\ m_1 & m_5 & -m_6 \end{pmatrix} \begin{pmatrix} j_4 & j_2 & j_6 \\ -m_4 & m_2 & m_6 \end{pmatrix} \begin{pmatrix} j_4 & j_5 & j_3 \\ m_4 & -m_5 & m_3 \end{pmatrix}$$

$$= \begin{Bmatrix} j_1 & j_2 & j_3 \\ j_4 & j_5 & j_6 \end{Bmatrix} \begin{pmatrix} j_1 & j_2 & j_3 \\ m_1 & m_2 & m_3 \end{pmatrix}, \tag{4.18}$$

WHICH IS A VARIANT OF EQUATION Z-4.15. *Hint:* START BY SHOWING THAT THE LEFT SIDE IS GIVEN BY

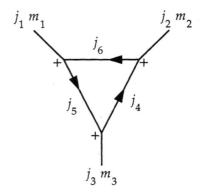

Next apply equation Z-4.55 to obtain the final result

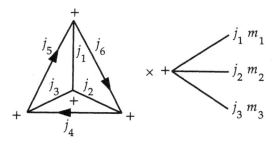

Let us start by finding a diagrammatic representation for the left side of equation 4.18:

$$\sum_{m_4 m_5 m_6} (-1)^{j_4+m_4+j_5+m_5+j_6+m_6} \begin{pmatrix} j_1 & j_5 & j_6 \\ m_1 & m_5 & -m_6 \end{pmatrix} \begin{pmatrix} j_4 & j_2 & j_6 \\ -m_4 & m_2 & m_6 \end{pmatrix} \begin{pmatrix} j_4 & j_5 & j_3 \\ m_4 & -m_5 & m_3 \end{pmatrix}$$

$$= \sum_{m_4 m_5 m_6} (-1)^{j_4+m_4+j_5+m_5+j_6+m_6}$$

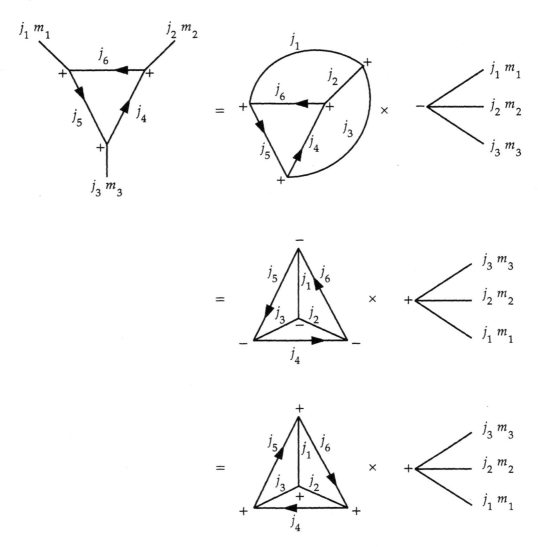

Applying equation Z-4.55 to equation 4.19, we obtain

$$= \left\{ \begin{array}{ccc} j_1 & j_2 & j_3 \\ j_4 & j_5 & j_6 \end{array} \right\} \left(\begin{array}{ccc} j_1 & j_2 & j_3 \\ m_1 & m_2 & m_3 \end{array} \right) \tag{4.20}$$

3. USE GRAPHICAL METHODS TO SHOW THAT THE SUM

$$\sum_{M_1 M_2 M_3 M_4} (-1)^{F_1 - M_1 + F_2 - M_2 + F_3 - M_3 + F_4 - M_4} \left(\begin{array}{ccc} F_1 & 1 & F_2 \\ -M_1 & \mu & M_2 \end{array} \right)$$

$$\times \left(\begin{array}{ccc} F_2 & 1 & F_3 \\ -M_2 & \mu' & M_3 \end{array} \right) \left(\begin{array}{ccc} F_3 & 1 & F_4 \\ -M_3 & \nu & M_4 \end{array} \right) \left(\begin{array}{ccc} F_4 & 1 & F_1 \\ -M_4 & \nu' & M_1 \end{array} \right) \tag{4.21}$$

CAN BE REEXPRESSED AS

$$\sum_{k,q} (-1)^{q + F_4 - F_1 + F_4 - F_3} (2k + 1) \left(\begin{array}{ccc} 1 & 1 & k \\ \mu & \mu' & q \end{array} \right) \left(\begin{array}{ccc} 1 & 1 & k \\ \nu & \nu' & -q \end{array} \right)$$

$$\times \left\{ \begin{array}{ccc} 1 & 1 & k \\ F_1 & F_3 & F_2 \end{array} \right\} \left\{ \begin{array}{ccc} 1 & 1 & k \\ F_1 & F_3 & F_4 \end{array} \right\}. \tag{4.22}$$

Hint: START BY SHOWING THAT THE SUM OVER THE FOUR 3-j SYMBOLS HAS THE GRAPH

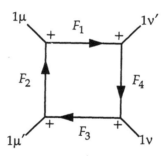

Equation 4.21 can be written as

$$\sum_{M_1 M_2 M_3 M_4} (-1)^{F_1 - M_1 + F_2 - M_2 + F_3 - M_3 + F_4 - M_4}$$

$$= \sum_{M_1 M_2 M_3 M_4}$$

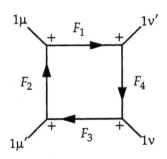

$$(4.23)$$

Applying equations Z-4.54 and Z-4.56 to equation 4.23 successively yields

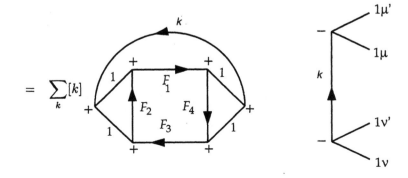

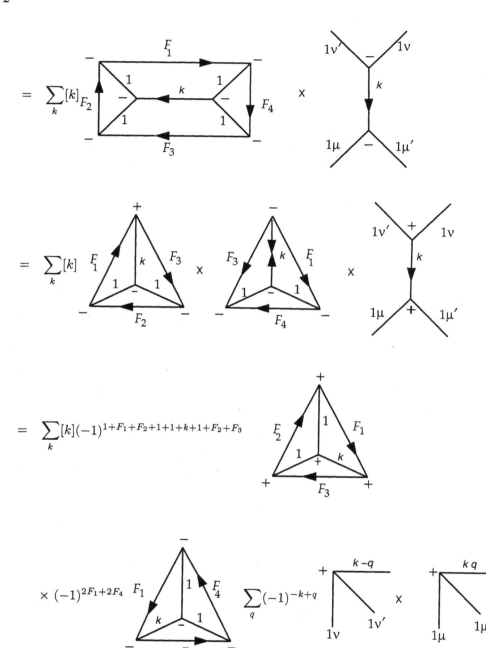

$$= \sum_k [k]_{F_2}$$

$$= \sum_k [k]$$

$$= \sum_k [k](-1)^{1+F_1+F_2+1+1+k+1+F_2+F_3}$$

$$\times (-1)^{2F_1+2F_4} \sum_q (-1)^{-k+q}$$

$$= \sum_{kq} (-1)^{q+3F_1+2F_2+F_3+2F_4}[k] \left\{ \begin{array}{ccc} 1 & k & 1 \\ F_3 & F_2 & F_1 \end{array} \right\} \left\{ \begin{array}{ccc} 1 & 1 & k \\ F_3 & F_1 & F_4 \end{array} \right\}$$

$$\times \begin{pmatrix} 1 & 1 & k \\ \nu & \nu' & -q \end{pmatrix} \begin{pmatrix} 1 & 1 & k \\ \mu & \mu' & q \end{pmatrix}. \tag{4.24}$$

Using the properties of 6-j symbols and the fact that $2F_2 + 2F_3 + 2$ should be even, we obtain equation 4.22, which is the desired result.

4.   USE GRAPHICAL METHODS TO PROVE THE BIEDENHARN-ELLIOT SUM RULE FOR 6-j SYMBOLS:

$$\begin{Bmatrix} j_1 & j_2 & j_3 \\ j_4 & j_5 & j_6 \end{Bmatrix} \begin{Bmatrix} j_1 & j_2 & j_3 \\ j_7 & j_8 & j_9 \end{Bmatrix} = \sum_j (-1)^{j_1+j_2+j_3+j_4+j_5+j_6+j_7+j_8+j_9+j}(2j+1)$$

$$\times \begin{Bmatrix} j_2 & j_4 & j_6 \\ j & j_9 & j_7 \end{Bmatrix} \begin{Bmatrix} j_1 & j_5 & j_6 \\ j & j_9 & j_8 \end{Bmatrix} \begin{Bmatrix} j_3 & j_4 & j_5 \\ j & j_8 & j_7 \end{Bmatrix}. \tag{4.25}$$

We express the product of the 6-j symbols graphically by writing

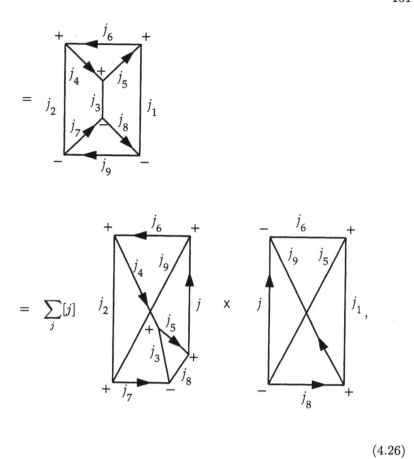

$$(4.26)$$

where we used equation Z-4.56. We evaluate each graph in equation 4.26 separately. The first graph becomes

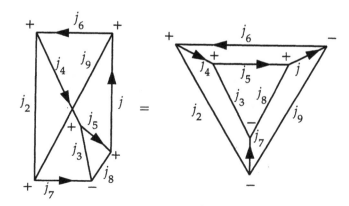

$=$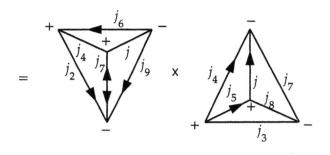

$=$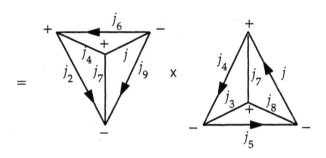

$= (-1)^{2j_2+2j_6+j_2+j_7+j_9+j+j_6+j_9}$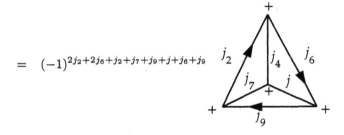

$\times (-1)^{j_4+j_7+j+j_3+j_7+j_8}$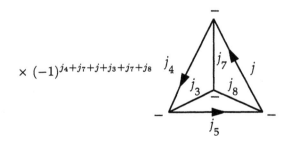

$$= (-1)^{3j_2+j_3+j_4+3j_6+3j_7+j_8+2j_9+2j} \left\{ \begin{matrix} j_4 & j & j_7 \\ j_9 & j_2 & j_6 \end{matrix} \right\} \left\{ \begin{matrix} j_7 & j_8 & j_3 \\ j_5 & j_4 & j \end{matrix} \right\}$$

$$= (-1)^{j_2+j_3+j_4+3j_6+j_7+j_8+2j} \left\{ \begin{matrix} j_4 & j_2 & j_6 \\ j_9 & j & j_7 \end{matrix} \right\} \left\{ \begin{matrix} j_5 & j_4 & j_3 \\ j_8 & j_7 & j \end{matrix} \right\}$$

$$= (-1)^{j_2+j_3+j_4+3j_6+j_7+j_8+2j} \left\{ \begin{matrix} j_2 & j_4 & j_6 \\ j & j_9 & j_7 \end{matrix} \right\} \left\{ \begin{matrix} j_3 & j_4 & j_5 \\ j & j_8 & j_7 \end{matrix} \right\}. \qquad (4.27)$$

The second graph can be written as

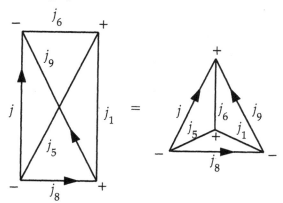

$$= (-1)^{2j_8+2j_9+j+j_5+j_8+j_1+j_9+j_8}$$

$$= (-1)^{j_1+j_5+3j_9+j} \left\{ \begin{matrix} j_6 & j_1 & j_5 \\ j_8 & j & j_9 \end{matrix} \right\}$$

$$= (-1)^{j_1+j_5+3j_9+j} \left\{ \begin{matrix} j_1 & j_5 & j_6 \\ j & j_9 & j_8 \end{matrix} \right\}. \qquad (4.28)$$

Finally, substituting equations 4.27 and 4.28 into 4.26, we get

$$\left\{ \begin{matrix} j_1 & j_2 & j_3 \\ j_4 & j_5 & j_6 \end{matrix} \right\} = \sum_j (-1)^{j_1+j_2+j_3+j_4+j_5+j_6+j_7+j_8+j_9+j}(2j+1)$$

$$\times \left\{ \begin{matrix} j_2 & j_4 & j_6 \\ j & j_9 & j_7 \end{matrix} \right\} \left\{ \begin{matrix} j_1 & j_5 & j_6 \\ j & j_9 & j_8 \end{matrix} \right\} \left\{ \begin{matrix} j_3 & j_4 & j_5 \\ j & j_8 & j_7 \end{matrix} \right\}. \qquad (4.29)$$

Chapter 5

SPHERICAL TENSOR OPERATORS

PROBLEM SET 3

1.   USE ANGULAR MOMENTUM COUPLING ALGEBRA TO PROVE THE VECTOR IDENTITY

$$\mathbf{a} \cdot (\mathbf{b} \times \mathbf{c}) = (\mathbf{a} \times \mathbf{b}) \cdot \mathbf{c}. \tag{5.1}$$

From the definition of a vector cross product, we know that

$$\mathbf{b} \times \mathbf{c} = (b_y c_z - b_z c_y)\hat{\mathbf{x}} + (b_z c_x - b_x c_z)\hat{\mathbf{y}} + (b_x c_y - b_y c_x)\hat{\mathbf{z}}. \tag{5.2}$$

The unit vectors in a spherical basis can be written as

$$\mathbf{e_1} = -2^{-1/2}(\hat{\mathbf{x}} + i\hat{\mathbf{y}}), \tag{5.3}$$

$$\mathbf{e_{-1}} = 2^{-1/2}(\hat{\mathbf{x}} - i\hat{\mathbf{y}}), \tag{5.4}$$

and

$$\mathbf{e_0} = \hat{\mathbf{z}}. \tag{5.5}$$

Taking Cartesian dot products, we obtain

$$(\mathbf{b} \times \mathbf{c}) \cdot \mathbf{e_1} = -2^{-1/2}(b_y c_z - b_z c_y) - i2^{-1/2}(b_z c_x - b_x c_z), \tag{5.6}$$

$$(\mathbf{b} \times \mathbf{c}) \cdot \mathbf{e_{-1}} = 2^{-1/2}(b_y c_z - b_z c_y) - i2^{-1/2}(b_z c_x - b_x c_z), \tag{5.7}$$

and

$$(\mathbf{b} \times \mathbf{c}) \cdot \mathbf{e_0} = b_x c_y - b_y c_x. \tag{5.8}$$

Comparing equations 5.6–5.8 with equations Z-5.41–Z-5.43, we observe that

$$(\mathbf{b} \times \mathbf{c}) \cdot \mathbf{e_q} = -i\sqrt{2}[b^{(1)} \otimes c^{(1)}]_q^{(1)}. \tag{5.9}$$

From equation Z-5.41, we know that

$$\mathbf{a} \cdot (\mathbf{b} \times \mathbf{c}) = -\sqrt{3} \left[a^{(1)} \otimes (\mathbf{b} \times \mathbf{c})^{(1)} \right]_0^{(0)}. \tag{5.10}$$

Comparing equations 5.9 and 5.10, we obtain

$$\mathbf{a} \cdot (\mathbf{b} \times \mathbf{c}) = i\sqrt{6} \left\{ a^{(1)} \otimes \left[b^{(1)} \otimes c^{(1)} \right]^{(1)} \right\}_0^{(0)}. \tag{5.11}$$

Using the recoupling transformation of three angular momenta given in equation Z-4.3, we obtain

$$\mathbf{a} \cdot (\mathbf{b} \times \mathbf{c}) = i\sqrt{6} \left\{ \left[a^{(1)} \otimes b^{(1)} \right]^{(1)} \otimes c^{(1)} \right\}_0^{(0)} \langle 110|110 \rangle, \tag{5.12}$$

where $\langle 110|110 \rangle$ is the recoupling coefficient. Because this coefficient has a value of 1, equation 5.12 becomes

$$\mathbf{a} \cdot (\mathbf{b} \times \mathbf{c}) = i\sqrt{6} \left\{ \left[a^{(1)} \otimes b^{(1)} \right]^{(1)} \otimes c^{(1)} \right\}_0^{(0)}. \tag{5.13}$$

Applying the same procedure to the right side of equation 5.1, we get

$$\mathbf{a} \cdot (\mathbf{b} \times \mathbf{c}) = (\mathbf{a} \times \mathbf{b}) \cdot \mathbf{c}. \tag{5.14}$$

2. DERIVE EQUATION Z-5.123 FROM EQUATION Z-5.122. OFFER AN INTERPRETATION WHY $C(J_i) \to 0$ AS $J_i \to \infty$ FOR THE RESONANCE FLUORESCENCE PROCESS INVOLVING $Q \uparrow$ OR $Q \downarrow$ BRANCHES.

The resonance fluorescence intensity for circularly polarized light can be derived from equations Z-5.105 and Z-5.106, using equations Z-5.120 and Z-5.121 for the polarization tensors of right and left circularly polarized light, respectively. As is the case for linear polarization, we must account for the different propagation directions of the excitation and detection light beams. Denoting the angle between the propagation directions of the two light beams as θ, we obtain

$$E_0^k(\hat{e}_a, \hat{e}_a^*) = \sum_q D_{q0}^k(0,\theta,0) \, E_q^k(\hat{e}_d, \hat{e}_d^*)$$

$$= D_{00}^k(0,\theta,0) \, E_0^k(\hat{e}_d, \hat{e}_d^*). \tag{5.15}$$

The degree of circular polarization, C, of the fluorescence is defined as

$$C = \frac{I_{\text{same}} - I_{\text{opposite}}}{I_{\text{same}} + I_{\text{opposite}}}. \tag{5.16}$$

For right polarized light, we insert equations Z-5.106, Z-5.120, and 5.15 into equation Z-5.105 and obtain for I_{same}

$$I_{\text{same}} \propto \left| \left\langle J_i || r^{(1)} || J_e \right\rangle \right|^2 \left| \left\langle J_e || r^{(1)} || J_f \right\rangle \right|^2 \sum_k (-1)(2k+1)^{\frac{1}{2}} \begin{pmatrix} 1 & 1 & k \\ 1 & -1 & 0 \end{pmatrix} D_{00}^k(0,\theta,0)$$

$$\times \quad (-1)(2k+1)^{\frac{1}{2}} \begin{pmatrix} 1 & 1 & k \\ 1 & -1 & 0 \end{pmatrix} a^k(J_i, J_e, J_f), \tag{5.17}$$

where we have used the fact that $q = 0$. An equivalent result is reached for left polarized light. Writing equation 5.17 explicitly, we obtain

$$I_{\text{same}} \propto \left| \left\langle J_i || r^{(1)} || J_e \right\rangle \right|^2 \left| \left\langle J_e || r^{(1)} || J_f \right\rangle \right|^2 \left[\begin{pmatrix} 1 & 1 & 0 \\ 1 & -1 & 0 \end{pmatrix}^2 a^0(J_i, J_e, J_f) D_{00}^0(0,\theta,0) \right.$$

$$+ 3 \begin{pmatrix} 1 & 1 & 1 \\ 1 & -1 & 0 \end{pmatrix}^2 a^1(J_i, J_e, J_f) D_{00}^1(0,\theta,0) + 5 \begin{pmatrix} 1 & 1 & 2 \\ 1 & -1 & 0 \end{pmatrix}^2 a^2(J_i, J_e, J_f) D_{00}^2(0,\theta,0) \right]$$

$$= \left| \left\langle J_i || r^{(1)} || J_e \right\rangle \right|^2 \left| \left\langle J_e || r^{(1)} || J_f \right\rangle \right|^2 \left[\frac{1}{3} a^0(J_i, J_e, J_f) P_0(\cos\theta) \right.$$

$$+ \frac{1}{2} a^1(J_i, J_e, J_f) P_1(\cos\theta) + \frac{1}{6} a^2(J_i, J_e, J_f) P_2(\cos\theta) \right]. \tag{5.18}$$

A similar calculation gives for I_{opposite}

$$I_{\text{opposite}} \quad \propto \quad \left| \left\langle J_{\text{i}} || r^{(1)} || J_{\text{e}} \right\rangle \right|^2 \; \left| \left\langle J_{\text{e}} || r^{(1)} || J_{\text{f}} \right\rangle \right|^2$$

$$\times \sum_k (2k+1) \begin{pmatrix} 1 & 1 & k \\ 1 & -1 & 0 \end{pmatrix} D_{00}^k(0,\theta,0) \begin{pmatrix} 1 & 1 & k \\ -1 & 1 & 0 \end{pmatrix} a^k(J_{\text{i}}, J_{\text{e}}, J_{\text{f}})$$

$$= \quad \left| \left\langle J_{\text{i}} || r^{(1)} || J_{\text{e}} \right\rangle \right|^2 \; \left| \left\langle J_{\text{e}} || r^{(1)} || J_{\text{f}} \right\rangle \right|^2 \left[\frac{1}{3} a^0(J_{\text{i}}, J_{\text{e}}, J_{\text{f}}) P_0(\cos\theta) \right.$$

$$\left. - \frac{1}{2} a^1(J_{\text{i}}, J_{\text{e}}, J_{\text{f}}) P_1(\cos\theta) + \frac{1}{6} a^2(J_{\text{i}}, J_{\text{e}}, J_{\text{f}}) P_2(\cos\theta) \right]. \tag{5.19}$$

From equations 5.18 and 5.19 we obtain

$$I_{\text{same}} - I_{\text{opposite}} = \left| \left\langle J_{\text{i}} || r^{(1)} || J_{\text{e}} \right\rangle \right|^2 \; \left| \left\langle J_{\text{e}} || r^{(1)} || J_{\text{f}} \right\rangle \right|^2 a^1(J_{\text{i}}, J_{\text{e}}, J_{\text{f}}) P_1(\cos\theta) \tag{5.20}$$

and

$$I_{\text{same}} + I_{\text{opposite}} = \left| \left\langle J_{\text{i}} || r^{(1)} || J_{\text{e}} \right\rangle \right|^2 \; \left| \left\langle J_{\text{e}} || r^{(1)} || J_{\text{f}} \right\rangle \right|^2$$

$$\times \left[\frac{2}{3} a^0(J_{\text{i}}, J_{\text{e}}, J_{\text{f}}) P_0(\cos\theta) + \frac{1}{3} a^2(J_{\text{i}}, J_{\text{e}}, J_{\text{f}}) P_2(\cos\theta) \right],$$

$$\tag{5.21}$$

which combine to yield

$$C = \frac{a^1(J_{\text{i}}, J_{\text{e}}, J_{\text{f}}) P_1(\cos\theta)}{\frac{2}{3} a^0(J_{\text{i}}, J_{\text{e}}, J_{\text{f}}) P_0(\cos\theta) + \frac{1}{3} a^2(J_{\text{i}}, J_{\text{e}}, J_{\text{f}}) P_2(\cos\theta)}$$

$$= \frac{\begin{Bmatrix} 1 & 1 & 1 \\ J_{\text{e}} & J_{\text{e}} & J_{\text{i}} \end{Bmatrix} \begin{Bmatrix} 1 & 1 & 1 \\ J_{\text{e}} & J_{\text{e}} & J_{\text{f}} \end{Bmatrix} P_1(\cos\theta)}{\frac{2}{3} \begin{Bmatrix} 1 & 1 & 0 \\ J_{\text{e}} & J_{\text{e}} & J_{\text{i}} \end{Bmatrix} \begin{Bmatrix} 1 & 1 & 0 \\ J_{\text{e}} & J_{\text{e}} & J_{\text{f}} \end{Bmatrix} + \frac{1}{3} \begin{Bmatrix} 1 & 1 & 2 \\ J_{\text{e}} & J_{\text{e}} & J_{\text{i}} \end{Bmatrix} \begin{Bmatrix} 1 & 1 & 2 \\ J_{\text{e}} & J_{\text{e}} & J_{\text{f}} \end{Bmatrix} P_2(\cos\theta)}. \tag{5.22}$$

A physical explanation of why C goes to zero for large J is based on the fact that in the classical limit the transition dipole moment μ is parallel to $\mathbf{J}$ and perpendicular to the internuclear axis for a Q-branch transition. Because $\mathbf{J}$ is a constant of the motion, it follows that for $Q(\uparrow)$, $Q(\downarrow)$ the fluorescence is linearly polarized in the same direction as the electric vector $\mathbf{E}$ of the absorbed light. Because linear polarization is an equal mixture of left and right circular polarization, it follows that the fluorescence will be the same regardless of the helicity of the absorbed photon.

3. Explore an alternative derivation of equation Z-5.105, which expresses the resonance fluorescence intensity in the absence of an external field. Start with equation Z-5.93 and write

$$\hat{e}_a \cdot \mathbf{r} = \sum_\mu (-1)^\mu (e_a)_{-\mu} \ r^1_\mu$$

$$\hat{e}^*_a \cdot \mathbf{r} = \sum_{\mu'} (-1)^{\mu'} (e^*_a)_{-\mu'} \ r^1_{\mu'}$$

$$\hat{e}_d \cdot \mathbf{r} = \sum_\nu (-1)^\nu (e_d)_{-\nu} \ r^1_\nu$$

$$\hat{e}^*_d \cdot \mathbf{r} = \sum_{\nu'} (-1)^{\nu'} (e^*_d)_{-\nu'} \ r^1_{\nu'}. \tag{5.23}$$

Apply the Wigner-Eckart theorem to show that

$$I \ \propto \ \sum_{\mu,\mu',\nu,\nu'} (-1)^{\mu+\mu'+\nu+\nu'} (e_a)_{-\mu}(e^*_a)_{-\mu'}(e_d)_{-\nu}(e^*_d)_{-\nu'}$$

$$\times \left| \left\langle \alpha_e J_e || r^{(1)} || \alpha_i J_i \right\rangle \right|^2 \ \left| \left\langle \alpha_e J_e || r^{(1)} || \alpha_f J_f \right\rangle \right|^2 \ (-1)^{J_i - J_e + J_f - J_e} \ S(J_i, J_e, J_f; \mu, \mu', \nu, \nu'), \tag{5.24}$$

where

$$S(J_i, J_e, J_f; \mu, \mu', \nu, \nu') = \sum_{M_i, M_e, M_f, M'_e} (-1)^{J_e - M_e + J_i - M_i + J_e - M'_e + J_f - M_f} \begin{pmatrix} J_e & 1 & J_i \\ -M_e & \mu & M_i \end{pmatrix}$$

$$\times \begin{pmatrix} J_i & 1 & J_e \\ -M_i & \mu' & M'_e \end{pmatrix} \begin{pmatrix} J_e & 1 & J_f \\ -M'_e & \nu & M_f \end{pmatrix} \begin{pmatrix} J_f & 1 & J_e \\ -M_f & \nu' & M_e \end{pmatrix}. \tag{5.25}$$

Use the results of Problem Set 2, part 3 to reexpress the summation S as

$$S = \sum_{kq} (-1)^{q + J_f - J_e + J_f - J_e} (2k+1) \begin{pmatrix} 1 & 1 & k \\ \mu & \mu' & q \end{pmatrix} \begin{pmatrix} 1 & 1 & k \\ \nu & \nu' & -q \end{pmatrix}$$

$$\times \begin{Bmatrix} 1 & 1 & k \\ J_e & J_e & J_i \end{Bmatrix} \begin{Bmatrix} 1 & 1 & k \\ J_e & J_e & J_f \end{Bmatrix}. \tag{5.26}$$

Show that on substitution this yields the expression for the resonance fluorescence intensity given by equations Z-5.105 and Z-5.106.

The resonance fluorescence signal for a zero external field is given by equation Z-5.93,

$$I \quad \propto \quad \sum_{M_i, M_f, M_e, M_e'} \langle \alpha_e J_e M_e | \hat{e}_a \cdot \mathbf{r} | \alpha_i J_i M_i \rangle \ \langle \alpha_i J_i M_i | \hat{e}_a^* \cdot \mathbf{r} | \alpha_e J_e M_e' \rangle$$

$$\times \langle \alpha_e J_e M_e' | \hat{e}_d \cdot \mathbf{r} | \alpha_f J_f M_f \rangle \ \langle \alpha_f J_f M_f | \hat{e}_d^* \cdot \mathbf{r} | \alpha_e J_e M_e \rangle . \tag{5.27}$$

Inserting equation 5.23 into equation 5.27 yields

$$I \quad \propto \quad \sum_{M_i, M_f, M_e, M_e'} \sum_{\mu, \mu', \nu, \nu'} (-1)^{\mu + \mu' + \nu + \nu'}$$

$$\times (e_a)_{-\mu} \left\langle \alpha_e J_e M_e \left| r_\mu^1 \right| \alpha_i J_i M_i \right\rangle (e_a^*)_{-\mu'} \left\langle \alpha_i J_i M_i \left| r_{\mu'}^1 \right| \alpha_e J_e M_e' \right\rangle$$

$$\times (e_d)_{-\nu} \left\langle \alpha_e J_e M_e' \left| r_\nu^1 \right| \alpha_f J_f M_f \right\rangle (e_d^*)_{-\nu'} \left\langle \alpha_f J_f M_f \left| r_{\nu'}^1 \right| \alpha_e J_e M_e \right\rangle . \tag{5.28}$$

Applying the Wigner-Eckart theorem to each of the matrix elements in equation 5.28, we obtain

$$I \quad \propto \quad \sum_{M_i, M_f, M_e, M_e'} \sum_{\mu, \mu', \nu, \nu'} (-1)^{\mu + \mu' + \nu + \nu'} (e_a)_{-\mu} (e_a^*)_{-\mu'} (e_d)_{-\nu} (e_d^*)_{-\nu'} (-1)^{J_i - J_e + J_f - J_e}$$

$$\times (-1)^{J_e - M_e + J_i - M_i + J_e - M_e' + J_f - M_f} \left| \left\langle \alpha_e J_e || r^{(1)} || \alpha_i J_i \right\rangle \right|^2 \ \left| \left\langle \alpha_e J_e || r^{(1)} || \alpha_f J_f \right\rangle \right|^2$$

$$\times \begin{pmatrix} J_e & 1 & J_i \\ -M_e & \mu & M_i \end{pmatrix} \begin{pmatrix} J_i & 1 & J_e \\ -M_i & \mu' & M_e' \end{pmatrix} \begin{pmatrix} J_e & 1 & J_f \\ -M_e' & \nu & M_f \end{pmatrix} \begin{pmatrix} J_f & 1 & J_e \\ -M_f & \nu' & M_e \end{pmatrix}$$

$$= \quad \sum_{\mu, \mu', \nu, \nu'} (-1)^{\mu + \mu' + \nu + \nu' + J_i - J_e + J_f - J_e} (e_a)_{-\mu} (e_a^*)_{-\mu'} (e_d)_{-\nu} (e_d^*)_{-\nu'}$$

$$\times \left| \left\langle \alpha_e J_e || r^{(1)} || \alpha_i J_i \right\rangle \right|^2 \ \left| \left\langle \alpha_e J_e || r^{(1)} || \alpha_f J_f \right\rangle \right|^2 S(J_i, J_e, J_f; \mu, \mu', \nu, \nu'), \tag{5.29}$$

where $S(J_i, J_e, J_f; \mu, \mu', \nu, \nu')$ is given by equation 5.25, where the phase factors come from equations Z-5.35 (twice) and Z-5.34 (4 times). Next, we substitute equation 5.26 into the previous expression to obtain

$$I \quad \propto \quad \sum_{\mu, \mu', \nu, \nu'} \sum_{k, q} (-1)^{\mu + \mu' + \nu + \nu'} (e_a)_{-\mu} (e_a^*)_{-\mu'} (e_d)_{-\nu} (e_d^*)_{-\nu'} \left| \left\langle \alpha_e J_e || r^{(1)} || \alpha_i J_i \right\rangle \right|^2 \ \left| \left\langle \alpha_e J_e || r^{(1)} || \alpha_f J_f \right\rangle \right|^2$$

$$\times (-1)^{q + J_f - J_e + J_i - J_e} (2k + 1) \begin{pmatrix} 1 & 1 & k \\ \mu & \mu' & q \end{pmatrix} \begin{pmatrix} 1 & 1 & k \\ \nu & \nu' & -q \end{pmatrix} \begin{Bmatrix} 1 & 1 & k \\ J_e & J_e & J_i \end{Bmatrix} \begin{Bmatrix} 1 & 1 & k \\ J_e & J_e & J_f \end{Bmatrix}$$

$$= \left|\left\langle \alpha_e J_e || r^{(1)} || \alpha_i J_i \right\rangle\right|^2 \ \left|\left\langle \alpha_e J_e || r^{(1)} || \alpha_f J_f \right\rangle\right|^2$$

$$\times \sum_{kq} \left\{ \left[\sum_{\mu,\mu'} (-1)^q (2k+1)^{\frac{1}{2}} (e_a)_{-\mu} (e_a^*)_{-\mu'} \begin{pmatrix} 1 & 1 & k \\ \mu & \mu' & q \end{pmatrix} \right] \right.$$

$$\times \left[\sum_{\nu,\nu'} (-1)^{-q} (2k+1)^{\frac{1}{2}} (e_d)_{-\nu} (e_d^*)_{-\nu'} \begin{pmatrix} 1 & 1 & k \\ \nu & \nu' & -q \end{pmatrix} \right]$$

$$\left. \times \left[(-1)^{q+J_f-J_e+J_i-J_e} \begin{Bmatrix} 1 & 1 & k \\ J_e & J_e & J_i \end{Bmatrix} \begin{Bmatrix} 1 & 1 & k \\ J_e & J_e & J_f \end{Bmatrix} \right] \right\}. \tag{5.30}$$

In the derivation of equation 5.30, we used the fact that $(-1)^{\mu+\mu'+\nu+\nu'} = (-1)^{-2q} = 1$. The sum in the first bracket can be rewritten as

$$(-1)^q (2k+1)^{\frac{1}{2}} \sum_{\mu,\mu'} (e_a)_{-\mu} (e_a^*)_{-\mu'} \begin{pmatrix} 1 & 1 & k \\ \mu & \mu' & q \end{pmatrix}$$

$$= (-1)^q (2k+1)^{\frac{1}{2}} \sum_{\mu} (e_a)_{-\mu} (e_a^*)_{q+\mu} \begin{pmatrix} 1 & 1 & k \\ \mu & -\mu-q & q \end{pmatrix}$$

$$= (-1)^q (2k+1)^{\frac{1}{2}} \sum_{\mu} (e_a)_{\mu} (e_a^*)_{q-\mu} \begin{pmatrix} 1 & 1 & k \\ -\mu & \mu-q & q \end{pmatrix}$$

$$= (-1)^{q+k} (2k+1)^{\frac{1}{2}} \sum_{\mu} (e_a)_{\mu} (e_a^*)_{q-\mu} \begin{pmatrix} 1 & 1 & k \\ \mu & q-\mu & -q \end{pmatrix}$$

$$= (-1)^k E_q^k(e_a, e_a^*). \tag{5.31}$$

The sum in the second bracket similarly may be written as $(-1)^k E_{-q}^k(e_d, e_d^*)$. Making these substitutions, we can rewrite equation 5.30 as

$$I \ \propto \ \left|\left\langle \alpha_e J_e || r^{(1)} || \alpha_i J_i \right\rangle\right|^2 \left|\left\langle \alpha_e J_e || r^{(1)} || \alpha_f J_f \right\rangle\right|^2 \sum_{k,q} (-1)^q E_q^k(e_a, e_a^*) E_{-q}^k(e_d, e_d^*)(-1)^{J_f-J_e+J_i-J_e}$$

$$\times \begin{Bmatrix} 1 & 1 & k \\ J_e & J_e & J_i \end{Bmatrix} \begin{Bmatrix} 1 & 1 & k \\ J_e & J_e & J_f \end{Bmatrix},$$

$$= \; \left| \left\langle \alpha_e J_e || r^{(1)} || \alpha_i J_i \right\rangle \right|^2 \; \left| \left\langle \alpha_e J_e || r^{(1)} || \alpha_f J_f \right\rangle \right|^2$$

$$\times \sum_{k,q} (-1)^q E_q^k(\mathbf{e}_a, \mathbf{e}_a^*) E_{-q}^k(\mathbf{e}_d, \mathbf{e}_d^*) a^k(J_i, J_e, J_f), \tag{5.32}$$

where

$$a^k(J_i, J_e, J_f) \;=\; (-1)^{J_f + J_e + J_i + J_e} \left\{ \begin{array}{ccc} 1 & 1 & k \\ J_e & J_e & J_i \end{array} \right\} \left\{ \begin{array}{ccc} 1 & 1 & k \\ J_e & J_e & J_f \end{array} \right\}. \tag{5.33}$$

4.   A SECOND-RANK CARTESIAN TENSOR T_{AB} ($A = X, Y, Z$ AND $B = X, Y, Z$) HAS NINE INDEPENDENT COMPONENTS. DECOMPOSE T_{AB} INTO ITS IRREDUCIBLE SPHERICAL TENSOR COMPONENTS. SHOW THAT T_{AB} IS COMPOSED OF A ZERO-RANK IRREDUCIBLE TENSOR, PROPORTIONAL TO $T_{XX} + T_{YY} + T_{ZZ} = \mathrm{Tr}[T]$ AND HAVING ONE INDEPENDENT COMPONENT, A FIRST-RANK IRREDUCIBLE TENSOR, PROPORTIONAL TO THE ANTISYMMETRIC TENSOR $\frac{1}{2}(T_{AB} - T_{BA})$ AND HAVING THREE INDEPENDENT COMPONENTS, AND A SECOND-RANK IRREDUCIBLE TENSOR, PROPORTIONAL TO THE SYMMETRIC TRACELESS TENSOR $\frac{1}{2}(T_{AB} + T_{BA}) - \frac{1}{3}\delta_{AB}\mathrm{Tr}[T]$ AND HAVING FIVE INDEPENDENT COMPONENTS. PRESENT A GENERAL PROCEDURE FOR DECOMPOSING AN ARBITRARY NTH-RANK CARTESIAN TENSOR INTO ITS SPHERICAL IRREDUCIBLE COMPONENTS. *Hint*: ANY CARTESIAN TENSOR MAY BE WRITTEN IN A SPHERICAL BASIS USING THE CORRESPONDENCE BETWEEN FIRST-RANK CARTESIAN TENSORS, T_X, T_Y, AND T_Z AND FIRST-RANK SPHERICAL TENSORS $T_1^{(1)}$, $T_0^{(1)}$, AND $T_{-1}^{(1)}$:

$$\begin{pmatrix} T_1^{(1)} \\ T_0^{(1)} \\ T_{-1}^{(1)} \end{pmatrix} = \begin{pmatrix} -\frac{1}{\sqrt{2}} & -\frac{i}{\sqrt{2}} & 0 \\ 0 & 0 & 1 \\ \frac{1}{\sqrt{2}} & \frac{-i}{\sqrt{2}} & 0 \end{pmatrix} \begin{pmatrix} T_X \\ T_Y \\ T_Z \end{pmatrix}, \tag{5.34}$$

OR

$$T_\mu^{(1)} = \sum_A U_{\mu A} T_A, \tag{5.35}$$

WITH THE INVERSE RELATION

$$\begin{pmatrix} T_X \\ T_Y \\ T_Z \end{pmatrix} = \begin{pmatrix} -\frac{1}{\sqrt{2}} & 0 & \frac{1}{\sqrt{2}} \\ \frac{i}{\sqrt{2}} & 0 & \frac{i}{\sqrt{2}} \\ 0 & 1 & 0 \end{pmatrix} \begin{pmatrix} T_1^{(1)} \\ T_0^{(1)} \\ T_{-1}^{(1)} \end{pmatrix}, \tag{5.36}$$

OR

$$T_A^{(1)} = \sum_\mu (U^{-1})_{A\mu} T_\mu^{(1)} = \sum_\mu U_{\mu A}^* T_\mu^{(1)}. \tag{5.37}$$

THUS, THE AB ELEMENT OF A SECOND-RANK CARTESIAN TENSOR T_{AB} IS WRITTEN IN A SPHERICAL BASIS AS

$$T_{AB} = \sum_{\mu,\nu} U_{\mu A}^* U_{\nu B}^* T_\mu^{(1)} T_\nu^{(1)}. \tag{5.38}$$

THE PRODUCT $T_\mu^{(1)}T_\nu^{(1)}$ IS NOT IRREDUCIBLE BUT TRANSFORMS UNDER ROTATION ACCORDING TO

$$
\begin{aligned}
\mathbf{R}T_\mu^{(1)}T_\nu^{(1)}\mathbf{R}^{-1} &= \mathbf{R}T_\mu^{(1)}\mathbf{R}^{-1}\mathbf{R}T_\nu^{(1)}\mathbf{R}^{-1} \\[2em]
&= \sum_{\mu',\nu'} D_{\mu'\mu}^1(R)\, D_{\nu'\nu}^1(R)\, T_{\mu'}^{(1)}T_{\nu'}^{(1)} \\[2em]
&= \sum_{\mu',\nu'}\sum_j \langle 1\mu',1\nu'|jm'\rangle\langle 1\mu,1\nu|jm\rangle\, D_{m'm}^j(R)\, T_{\mu'}^{(1)}T_{\nu'}^{(1)},
\end{aligned}
\tag{5.39}
$$

AND IT FOLLOWS THAT

$$
\begin{aligned}
\mathbf{R}T_{AB}\mathbf{R}^{-1} &= \sum_j\sum_{\mu,\nu} U_{\mu A}^* U_{\nu B}^* \sum_{\mu',\nu'} \langle 1\mu',1\nu'|jm'\rangle\langle 1\mu,1\nu|jm\rangle D_{m'm}^j(R)\, T_{\mu'}^{(1)}T_{\nu'}^{(1)} \\[2em]
&= \sum_j\sum_{m,m'} C_{jm}^{AB} D_{m'm}^j(R)\, T_{m'}^{(j)}.
\end{aligned}
\tag{5.40}
$$

SOLVE FOR THE COEFFICIENTS C_{jm}^{AB} IN THE EXPANSION. THE GENERALIZATION TO HIGHER-ORDER CARTESIAN TENSORS RELIES ON THE REPEATED APPLICATION OF THE PROCEDURE SHOWN ABOVE.

We want to expand T_{AB} as a sum of irreducible spherical tensors. This expansion can be accomplished by evaluating equation 5.40 in the limit of zero rotation. Noting that $\mathrm{d}_{m'm}^j(\theta=0)=\delta_{m'm}$ (equation Z-3.71), we obtain for each value of j ($j=0,1,2$),

$$
\sum_{\mu,\nu} U_{\mu A}^* U_{\nu B}^* \langle 1\mu,1\nu|jm\rangle \left(\sum_{\mu',\nu'} \langle 1\mu',1\nu'|jm\rangle T_{\mu'}^{(1)}T_{\nu'}^{(1)} \right) = \sum_m C_{jm}^{AB}\, T_m^{(j)}.
\tag{5.41}
$$

The sum in parenthesis is a contraction of first-order tensors (equation Z-5.36), which may be evaluated to give

$$
\sum_{\mu,\nu} U_{\mu A}^* U_{\nu B}^* \langle 1\mu,1\nu|jm\rangle T_m^{(j)} = \sum_m C_{jm}^{AB}\, T_m^{(j)}.
\tag{5.42}
$$

Equating the coefficients of $T_m^{(j)}$, we obtain

$$
C_{jm}^{AB} = \sum_{\mu,\nu} \langle 1\mu,1\nu|jm\rangle U_{\mu A}^* U_{\nu B}^*.
\tag{5.43}
$$

Using Table Z-2.4 to evaluate the Clebsch-Gordan coefficients, we obtain

$$
C_{00}^{AB} = \frac{1}{\sqrt{3}}\left(U_{1A}^* U_{-1B}^* - U_{0A}^* U_{0B}^* + U_{-1A}^* U_{1B}^* \right),
\tag{5.44}
$$

$$
C_{11}^{AB} = \frac{1}{\sqrt{2}}\left(U_{1A}^* U_{0B}^* - U_{0A}^* U_{1B}^* \right),
\tag{5.45}
$$

$$C_{10}^{AB} = \frac{1}{\sqrt{2}} \left(U_{1A}^* U_{-1B}^* - U_{-1A}^* U_{1B}^* \right), \tag{5.46}$$

$$C_{1-1}^{AB} = \frac{1}{\sqrt{2}} \left(-U_{-1A}^* U_{0B}^* + U_{0A}^* U_{-1B}^* \right), \tag{5.47}$$

$$C_{22}^{AB} = U_{1A}^* U_{1B}^*, \tag{5.48}$$

$$C_{21}^{AB} = \frac{1}{\sqrt{2}} \left(U_{1A}^* U_{0B}^* + U_{0A}^* U_{1B}^* \right), \tag{5.49}$$

$$C_{20}^{AB} = \frac{1}{\sqrt{6}} U_{1A}^* U_{-1B}^* + \sqrt{\frac{2}{3}} U_{0A}^* U_{0B}^* + \frac{1}{\sqrt{6}} U_{-1A}^* U_{1B}^*, \tag{5.50}$$

$$C_{2-1}^{AB} = \frac{1}{\sqrt{2}} \left(U_{-1A}^* U_{0B}^* + U_{0A}^* U_{-1B}^* \right), \tag{5.51}$$

and

$$C_{2-2}^{AB} = U_{-1A}^* U_{-1B}^*. \tag{5.52}$$

We illustrate the calculation of $C_{jm}^{AB}\, T_m^{(j)}$ for a number of terms, using equations Z-5.41 to Z-5.44 to evaluate $T_m^{(j)}$ and taking matrix elements $U_{\mu A}^*$ and $U_{\nu B}^*$ directly from equation 5.34:

$$
\begin{aligned}
C_{00}^{xx} &= \frac{1}{\sqrt{3}} \left[\left(-\frac{1}{\sqrt{2}} \right) \left(\frac{1}{\sqrt{2}} \right) + 0 + \left(\frac{1}{\sqrt{2}} \right) \left(-\frac{1}{\sqrt{2}} \right) \right] \\
&= -\frac{1}{\sqrt{3}},
\end{aligned}
\tag{5.53}
$$

$$
\begin{aligned}
C_{00}^{yy} &= C_{00}^{zz} \\
&= -\frac{1}{\sqrt{3}},
\end{aligned}
\tag{5.54}
$$

$$C_{00}^{AB} = 0 \qquad\qquad \text{for } A \neq B, \tag{5.55}$$

$$C_{00}^{xx}\, T_0^{(0)} = \frac{1}{3} \left(A_x B_x + A_y B_y + A_z B_z \right), \tag{5.56}$$

$$
\begin{aligned}
C_{11}^{xy} &= C_{1-1}^{xy} \\
&= 0,
\end{aligned}
\tag{5.57}
$$

$$
\begin{aligned}
C_{10}^{xy} &= \frac{1}{\sqrt{2}} \left[\left(-\frac{1}{\sqrt{2}} \right) \left(\frac{i}{\sqrt{2}} \right) - \left(\frac{1}{\sqrt{2}} \right) \left(\frac{i}{\sqrt{2}} \right) \right] \\
&= -\frac{i}{\sqrt{2}},
\end{aligned}
\tag{5.58}
$$

$$C_{10}^{xy}\, T_0^{(1)} = \frac{1}{2} \left(A_x B_y - A_y B_x \right), \tag{5.59}$$

$$C_{22}^{xy} = -C_{2-2}^{xy}$$
$$= -\frac{i}{2},$$
(5.60)

$$C_{21}^{xy} = -C_{2-1}^{xy}$$
$$= 0,$$
(5.61)

$$C_{20}^{xy} = \frac{1}{\sqrt{6}}\left(-\frac{1}{\sqrt{2}}\right)\left(\frac{i}{\sqrt{2}}\right) + \frac{1}{\sqrt{6}}\left(\frac{1}{\sqrt{2}}\right)\left(\frac{i}{\sqrt{2}}\right) + \sqrt{\frac{2}{3}} \times 0$$
$$= 0,$$
(5.62)

and

$$C_{22}^{xy}T_2^{(2)} + C_{2-2}^{xy}T_{-2}^{(2)} = A_x B_y + A_y B_x.$$
(5.63)

Summing equation 5.41 over j, we obtain for the Cartesian tensor

$$T_{AB} = \sum_{j=0}^{2} \sum_{m=-j}^{j} C_{jm}^{AB} T_m^{(j)}.$$
(5.64)

Collecting all of the terms in equation 5.64, we find that

$$T_{AB} = \frac{1}{3}\mathcal{T}\delta_{AB} + \mathcal{A}_C + \mathcal{S}_{AB},$$
(5.65)

where $\mathcal{T}$ is the trace of T (equation 5.56), $\mathcal{A}$ is an antisymmetric tensor (equation 5.59) with components

$$\mathcal{A}_C = \frac{1}{2}\left(T_{AB} - T_{BA}\right), \qquad A, B, C \text{ are cyclic},$$
(5.66)

and $\mathcal{S}$ is a symmetric tensor (equation 5.63) with components

$$\mathcal{S}_{AB} = \frac{1}{2}\left(T_{AB} + T_{BA} - \frac{2}{3}\mathcal{T}\delta_{AB}\right).$$
(5.67)

This solution may be generalized to higher-order tensors by writing the nth-order Cartesian tensor as a product of first-order Cartesian tensors,

$$T_{ABCD\ldots} = \sum_{\alpha,\beta,\gamma,\delta,\ldots} \left(U_{\alpha A}^* U_{\beta B}^* U_{\gamma C}^* U_{\delta D}^* \ldots\right)\left(T_\alpha^{(1)} T_\beta^{(1)} T_\gamma^{(1)} T_\delta^{(1)}\ldots\right).$$
(5.68)

Rotation of the first-order tensors gives

$$\mathbf{R}T_\alpha^{(1)}T_\beta^{(1)}T_\gamma^{(1)}T_\delta^{(1)}\ldots\mathbf{R}^{-1} = \mathbf{R}T_\alpha^{(1)}\mathbf{R}^{-1}\mathbf{R}T_\beta^{(1)}\mathbf{R}^{-1}\mathbf{R}T_\gamma^{(1)}\mathbf{R}^{-1}\mathbf{R}T_\delta^{(1)}\mathbf{R}^{-1}\ldots$$

$$= \sum_{\alpha',\beta',\gamma',\delta',\ldots} D_{\alpha'\alpha}^1(R)\, D_{\beta'\beta}^1(R)\, D_{\gamma'\gamma}^1(R) D_{\delta'\delta}^1(R)\ldots T_{\alpha'}^{(1)} T_{\beta'}^{(1)} T_{\gamma'}^{(1)} T_{\delta'}^{(1)} \ldots$$

$$= \sum_{\alpha',\beta',\gamma',\delta',\ldots}\sum_{j,j',\ldots} \langle 1\alpha', 1\beta'|j\mu\rangle \langle 1\alpha, 1\beta|jm\rangle D_{\mu m}^j$$

$$\times \langle 1\gamma', 1\delta'|j'\mu'\rangle \langle 1\gamma, 1\delta|j'm'\rangle D_{\mu'm'}^{j'}\ldots T_{\alpha'}^{(1)} T_{\beta'}^{(1)} T_{\gamma'}^{(1)} T_{\delta'}^{(1)} \ldots$$
(5.69)

Successive contraction of the first-order tensors leads to the desired generalization.

APPLICATION 11
THE ENERGY LEVELS OF A TWO-VALENCE-ELECTRON ATOM REVISITED

A.

WORK OUT THE TERM ENERGIES FOR THE p^2 CONFIGURATION, USING SPHERICAL TENSOR METHODS.

Recalling Application 4, we know that to find the relative term energies, we need to calculate the matrix elements of the Coulomb repulsion term $\frac{e^2}{r_{12}}$ for a given configuration for states of the same L and S,

$$E\left({}^{2S+1}L\right) = \left\langle \ell_1 = 1,\ \ell_2 = 1\ SM_S, LM_L \left| \frac{e^2}{r_{12}} \right| \ell_1 = 1,\ \ell_2 = 1\ SM_S, LM_L \right\rangle. \tag{5.70}$$

We start by expanding $\frac{e^2}{r_{12}}$ as a series of Legendre polynomials, using equation Z-3.4.13,

$$\frac{e^2}{r_{12}} = e^2 \sum_k \frac{r_<^k}{r_>^{k+1}} P_k(\cos\Theta). \tag{5.71}$$

Writing the Legendre polynomials as a scalar contraction of tensors, using equation Z-5.55, we get

$$\frac{e^2}{r_{12}} = e^2 \sum_k \frac{r_<^k}{r_>^{k+1}} C^k(1) \cdot C^k(2). \tag{5.72}$$

Substituting equation 5.72 into equation 5.70 and using equation Z-3.4.17, we obtain

$$E\left({}^{2S+1}L\right) = \sum_k \langle p^2 SM_S, LM_L \left| C^k(1) \cdot C^k(2) \right| p^2 SM_S, LM_L \rangle$$

$$\times e^2 \int_0^\infty r_1^2 dr_1 \int_0^\infty r_2^2 dr_2 \frac{r_<^k}{r_>^{k+1}} [R_p(1)R_p(2)]^2$$

$$= \sum_k \langle p^2 SM_S, LM_L \left| C^k(1) \cdot C^k(2) \right| p^2 SM_S, LM_L \rangle \ F^k. \tag{5.73}$$

Applying equation Z-5.71 to equation 5.73, we obtain

$$E\left({}^{2S+1}L\right) = \sum_k (-1)^L \left\{ \begin{matrix} 1 & 1 & L \\ 1 & 1 & k \end{matrix} \right\} \langle 1 \left\| C^k \right\| 1 \rangle^2 F^k. \tag{5.74}$$

The reduced matrix element in equation 5.74 can be evaluated using equation Z-5.30,

$$\langle \ell_1 \left\| C^k \right\| \ell_3 \rangle = (-1)^{\ell_1} [(2\ell_1 + 1)(2\ell_3 + 1)]^{\frac{1}{2}} \left(\begin{matrix} \ell_1 & k & \ell_3 \\ 0 & 0 & 0 \end{matrix} \right). \tag{5.75}$$

We are now in a position to evaluate the individual term energies.

For the ^{1}S term, we get

$$E(^1S) = \sum_k \left\{ \begin{array}{ccc} 1 & 1 & 0 \\ 1 & 1 & k \end{array} \right\} \langle 1 \| C^k \| 1 \rangle^2 F^k$$

$$= 9 \sum_k \left\{ \begin{array}{ccc} 1 & 1 & 0 \\ 1 & 1 & k \end{array} \right\} \left(\begin{array}{ccc} 1 & k & 1 \\ 0 & 0 & 0 \end{array} \right)^2 F^k$$

$$= 9 \left[\frac{1}{3} \left\{ \begin{array}{ccc} 1 & 1 & 0 \\ 1 & 1 & 0 \end{array} \right\} F^0 + \frac{2}{15} \left\{ \begin{array}{ccc} 1 & 1 & 0 \\ 1 & 1 & 2 \end{array} \right\} F^2 \right]$$

$$= F^0 + \frac{2}{5} F^2. \tag{5.76}$$

For the ^{3}P term, we get

$$E(^3P) = -\sum_k \left\{ \begin{array}{ccc} 1 & 1 & 1 \\ 1 & 1 & k \end{array} \right\} \left(\begin{array}{ccc} 1 & k & 1 \\ 0 & 0 & 0 \end{array} \right)^2 F^k$$

$$= -9 \left[\frac{1}{3} \left\{ \begin{array}{ccc} 1 & 1 & 1 \\ 1 & 1 & 0 \end{array} \right\} F^0 + \frac{2}{15} \left\{ \begin{array}{ccc} 1 & 1 & 1 \\ 1 & 1 & 2 \end{array} \right\} F^2 \right]$$

$$= F^0 - \frac{1}{5} F^2. \tag{5.77}$$

For the ^{1}D term, we get

$$E(^1D) = \sum_k \left\{ \begin{array}{ccc} 1 & 1 & 2 \\ 1 & 1 & k \end{array} \right\} \left(\begin{array}{ccc} 1 & k & 1 \\ 0 & 0 & 0 \end{array} \right)^2 F^k$$

$$= 9 \left[\frac{1}{3} \left\{ \begin{array}{ccc} 1 & 1 & 2 \\ 1 & 1 & 0 \end{array} \right\} F^0 + \frac{2}{15} \left\{ \begin{array}{ccc} 1 & 1 & 2 \\ 1 & 1 & 2 \end{array} \right\} F^2 \right]$$

$$= F^0 + \frac{1}{25} F^2. \tag{5.78}$$

To evaluate the 3-j and 6-j symbols, we used equations Z-4.9 and Z-4.19 and Tables Z-2.5 and Z-4.1. Using the definitions $F_0 = F^0$ and $F_2 = \frac{1}{25} F^2$, we obtain

$$E(^1S) = F_0 + 10F_2, \tag{5.79}$$

$$E(^3P) = F_0 - 5F_2, \tag{5.80}$$

and

$$E(^1D) = F_0 + F_2. \tag{5.81}$$

B.

BY CONSIDERING ONLY THE ELECTROSTATIC INTERACTION, THE RESULTING ENERGY EXPRESSION DOES NOT CONTAIN J, AND HENCE THE DIFFERENT J LEVELS $^{2S+1}L_J$ OF THE TERM ^{2S+1}L HAVE THE SAME ENERGY. HOWEVER, INCLUSION OF THE SPIN-ORBIT INTERACTION TERM

$$H_{SO} = \sum_i \xi(r_i)\boldsymbol{\ell}_i \cdot \boldsymbol{s}_i$$

IN THE HAMILTONIAN REMOVES THIS DEGENERACY. FIND $E(^{2S+1}L_J)$, TREATING H_{SO} AS A PERTURBATION. SET UP, BUT DO NOT EVALUATE, THE EXACT EXPRESSIONS FOR $E(^{2S+1}L_J)$ WHEN THE SPIN-ORBIT INTERACTION IS NOT SMALL COMPARED WITH THE SEPARATION BETWEEN TERM ENERGIES. EXPRESS THE ENERGY LEVELS IN TERMS OF THE SPIN-ORBIT PARAMETER

$$\zeta_{n\ell} = \int R_{n\ell}(r)\xi(r)R_{n\ell}(r)r^2 dr.$$

From equation Z-5.79, we see that the spin-orbit coupling term is given by

$$\left\langle {}^{2S+1}L_J M_J \left| H_{SO} \right| {}^{2S'+1}L'_{J'} M'_{J'} \right\rangle = \zeta_{n_1\ell_1;n'_1\ell'_1}\delta_{JJ'}\delta_{M_J M'_{J'}}\delta_{\ell_1\ell'_1}\delta_{\ell_2\ell'_2}(-1)^{S+S'+J+\ell_1+\ell_2+1}$$

$$\times \left[(2L'+1)(2L+1)(2S'+1)(2S+1)(2\ell_1+1)\ell_1(\ell_1+1)\frac{3}{2}\right]^{\frac{1}{2}}$$

$$\times \left\{\begin{matrix} L & S & J \\ S' & L' & 1 \end{matrix}\right\}\left\{\begin{matrix} \ell_1 & L & \ell_2 \\ L' & \ell_1 & 1 \end{matrix}\right\}\left\{\begin{matrix} \frac{1}{2} & S & \frac{1}{2} \\ S' & \frac{1}{2} & 1 \end{matrix}\right\}$$

$$+ \zeta_{n_2\ell_2;n'_2\ell'_2}\delta_{JJ'}\delta_{M_J M'_{J'}}\delta_{\ell_1\ell'_1}\delta_{\ell_2\ell'_2}(-1)^{2S+L+L'+J+\ell_1+\ell_2+1}$$

$$\times \left[(2L'+1)(2L+1)(2S'+1)(2S+1)(2\ell_2+1)\ell_2(\ell_2+1)\frac{3}{2}\right]^{\frac{1}{2}}$$

$$\times \left\{\begin{matrix} L & S & J \\ S' & L' & 1 \end{matrix}\right\}\left\{\begin{matrix} \ell_2 & L & \ell_1 \\ L' & \ell_2 & 1 \end{matrix}\right\}\left\{\begin{matrix} \frac{1}{2} & S & \frac{1}{2} \\ S' & \frac{1}{2} & 1 \end{matrix}\right\}. \tag{5.82}$$

Working in the first-order perturbation regime requires evaluation of only the diagonal terms. For a configuration with two equivalent electrons, np^2 ($\ell = 1$), the diagonal terms have the form

$$\left\langle {}^{2S+1}L_J M_J \left| H_{SO} \right| {}^{2S+1}L_J M_J \right\rangle = 6(2L+1)(2S+1)\,\zeta_{np}\,(-1)^{2S+J+1}$$

$$\times \left\{\begin{matrix} L & S & J \\ S & L & 1 \end{matrix}\right\}\left\{\begin{matrix} 1 & L & 1 \\ L & 1 & 1 \end{matrix}\right\}\left\{\begin{matrix} \frac{1}{2} & S & \frac{1}{2} \\ S & \frac{1}{2} & 1 \end{matrix}\right\}. \tag{5.83}$$

The next step is the evaluation of equation 5.83 for each L, S, J combination.

i) $L = 0$, $S = 0$, $J = 0$

$$\langle {}^1S_0\, 0 | H_{SO} | {}^1S_0\, 0 \rangle = -6\, \zeta_{np} \begin{Bmatrix} 0 & 0 & 0 \\ 0 & 0 & 1 \end{Bmatrix} \begin{Bmatrix} 1 & 0 & 1 \\ 0 & 1 & 1 \end{Bmatrix} \begin{Bmatrix} \frac{1}{2} & 0 & \frac{1}{2} \\ 0 & \frac{1}{2} & 1 \end{Bmatrix}$$

$$= 0. \tag{5.84}$$

ii) $L = 1$, $S = 1$, $J = 0$

$$\langle {}^3P_0\, 0 | H_{SO} | {}^3P_0\, 0 \rangle = -54\, \zeta_{np} \begin{Bmatrix} 1 & 1 & 0 \\ 1 & 1 & 1 \end{Bmatrix} \begin{Bmatrix} 1 & 1 & 1 \\ 1 & 1 & 1 \end{Bmatrix} \begin{Bmatrix} \frac{1}{2} & 1 & \frac{1}{2} \\ 1 & \frac{1}{2} & 1 \end{Bmatrix}$$

$$= -54\zeta_{np} \left(-\frac{1}{3}\right)\left(\frac{1}{6}\right)\left(-\frac{1}{3}\right)$$

$$= -\zeta_{np}. \tag{5.85}$$

iii) $L = 1$, $S = 1$, $J = 1$

$$\langle {}^3P_1\, M | H_{SO} | {}^3P_1\, M \rangle = 54\, \zeta_{np} \begin{Bmatrix} 1 & 1 & 1 \\ 1 & 1 & 1 \end{Bmatrix} \begin{Bmatrix} 1 & 1 & 1 \\ 1 & 1 & 1 \end{Bmatrix} \begin{Bmatrix} \frac{1}{2} & 1 & \frac{1}{2} \\ 1 & \frac{1}{2} & 1 \end{Bmatrix}$$

$$= 54\zeta_{np} \left(\frac{1}{6}\right)^2\left(-\frac{1}{3}\right)$$

$$= -\frac{1}{2}\zeta_{np}. \tag{5.86}$$

iv) $L = 1$, $S = 1$, $J = 2$

$$\langle {}^3P_2\, M | H_{SO} | {}^3P_2\, M \rangle = -54\, \zeta_{np} \begin{Bmatrix} 1 & 1 & 2 \\ 1 & 1 & 1 \end{Bmatrix} \begin{Bmatrix} 1 & 1 & 1 \\ 1 & 1 & 1 \end{Bmatrix} \begin{Bmatrix} \frac{1}{2} & 1 & \frac{1}{2} \\ 1 & \frac{1}{2} & 1 \end{Bmatrix}$$

$$= -54\zeta_{np} \left(\frac{1}{6}\right)\left(\frac{1}{6}\right)\left(-\frac{1}{3}\right)$$

$$= \frac{1}{2}\zeta_{np}. \tag{5.87}$$

v)   L = 2, S = 0, J = 2

$$\langle {}^1D_2\, M\,|H_{SO}|\,{}^1D_2\, M\rangle \;=\; -54\,\zeta_{np}\begin{Bmatrix} 2 & 0 & 2 \\ 0 & 2 & 1 \end{Bmatrix}\begin{Bmatrix} 1 & 2 & 1 \\ 2 & 1 & 1 \end{Bmatrix}\begin{Bmatrix} \tfrac{1}{2} & 0 & \tfrac{1}{2} \\ 0 & \tfrac{1}{2} & 1 \end{Bmatrix}$$

$$=\;\; 0. \hspace{7cm} (5.88)$$

In evaluating equation 5.83, we used the symmetry properties of the 6-j symbols, including the special case of a 6-j symbol with a vanishing argument. Numerical values of the 6-j symbols were obtained from Table Z-4.1. Combining these results with the values obtained in part **A**, we obtain the energies in the first-order perturbation limit:

$$E({}^1S_0) \;=\; F_0 + 10F_2,$$

$$E({}^3P_0) \;=\; F_0 - 5F_2 - \zeta_{np},$$

$$E({}^3P_1) \;=\; F_0 - 5F_2 - \frac{1}{2}\zeta_{np},$$

$$E({}^3P_2) \;=\; F_0 - 5F_2 + \frac{1}{2}\zeta_{np},$$

$$E({}^1D_2) \;=\; F_0 + F_2. \hspace{6cm} (5.89)$$

To obtain the exact energies, we need to consider also the nondiagonal matrix elements. Equation 5.82 shows that H_{SO} connects states of only the same total angular momentum J and projection M_J. It also shows that the energy does not depend on M_J, leading to a degeneracy of the magnetic states that cannot be broken by spin-orbit interaction. According to equation 5.82, mixing occurs between ${}^1S_0$ and ${}^3P_0$ and between ${}^3P_2$ and ${}^1D_2$. The state corresponding to the ${}^3P_1$ term is not affected by this type of interaction.

${}^1S_0$ and ${}^3P_0$

We start by setting up the secular determinant

$$\begin{vmatrix} \langle {}^1S_0\,|H|\,{}^1S_0\rangle - E & \langle {}^1S_0\,|H|\,{}^3P_0\rangle \\[2mm] \langle {}^3P_0\,|H|\,{}^1S_0\rangle & \langle {}^3P_0\,|H|\,{}^3P_0\rangle - E \end{vmatrix} \;=\; 0, \hspace{3cm} (5.90)$$

where $H = e^2/r_{12} + H_{SO}$. The diagonal terms for H_{SO} have already been evaluated (equations 5.89), leaving only the off-diagonal contributions. By setting $L = 0, S = 0, L' = 1, S' = 1, J = 0$ in equation 5.82 and using Table Z-4.1, we obtain

$$\langle ^1S_0 |H_{SO}| ^3P_0\rangle = \langle ^3P_0 |H_{SO}| ^1S_0\rangle$$

$$= 18\,\zeta_{np} \left\{ \begin{array}{ccc} 0 & 0 & 0 \\ 1 & 1 & 1 \end{array} \right\} \left\{ \begin{array}{ccc} 1 & 0 & 1 \\ 1 & 1 & 1 \end{array} \right\} \left\{ \begin{array}{ccc} \frac{1}{2} & 0 & \frac{1}{2} \\ 1 & \frac{1}{2} & 1 \end{array} \right\}$$

$$= 18\zeta_{np} \left(\frac{1}{\sqrt{3}} \right) \left(-\frac{1}{3} \right) \left(\frac{1}{\sqrt{6}} \right)$$

$$= -\sqrt{2}\,\zeta_{np}. \tag{5.91}$$

Substituting equations 5.89 and 5.91 into the secular determinant and solving for the energies, we obtain

$$0 = (F_0 + 10F_2 - E)(F_0 - 5F_2 - \zeta_{np} - E) - 2\zeta_{np}^2$$

$$= (E - F_0)^2 - (5F_2 - \zeta_{np})(E - F_0)$$

$$-10F_2(5F_2 + \zeta_{np}) - 2\zeta_{np}^2, \tag{5.92}$$

and thus

$$E - F_0 = \frac{1}{2}(5F_2 - \zeta_{np}) \pm \frac{1}{2} \left[(5F_2 - \zeta_{np})^2 + 40F_2(5F_2 + \zeta_{np}) + 8\zeta_{np}^2 \right]^{1/2}. \tag{5.93}$$

The two solutions can be assigned to the two states by looking at their behavior as $\zeta_{np} \to 0$ (i.e., in the first-order perturbation limit), giving

$$E(^1S_0) = F_0 + \frac{1}{2}(5F_2 - \zeta_{np}) + \frac{1}{2} \left[225F_2^2 + 30F_2\zeta_{np} + 9\zeta_{np}^2 \right]^{\frac{1}{2}} \tag{5.94}$$

$$E(^3P_0) = F_0 + \frac{1}{2}(5F_2 - \zeta_{np}) - \frac{1}{2} \left[225F_2^2 + 30F_2\zeta_{np} + 9\zeta_{np}^2 \right]^{\frac{1}{2}}. \tag{5.95}$$

$^3P_2$ and $^1D_2$

The diagonal terms of the secular determinant are again given by equation 5.89. The off-diagonal terms of H_{SO} can be obtained by setting $L = 1, S = 1, L' = 2, S' = 0, J = 2$ in equation 5.82,

$$\langle ^3P_2 |H_{SO}| ^1D_2\rangle = \langle ^1D_2 |H_{SO}| ^3P_2\rangle$$

$$= 2\,\zeta_{np} \times 9\sqrt{5} \left\{ \begin{array}{ccc} 1 & 1 & 2 \\ 0 & 2 & 1 \end{array} \right\} \left\{ \begin{array}{ccc} 1 & 1 & 1 \\ 2 & 1 & 1 \end{array} \right\} \left\{ \begin{array}{ccc} \frac{1}{2} & 1 & \frac{1}{2} \\ 0 & \frac{1}{2} & 1 \end{array} \right\}$$

$$= 18\sqrt{5}\zeta_{np} \left(\frac{1}{\sqrt{15}} \right) \left(\frac{1}{6} \right) \left(\frac{1}{\sqrt{6}} \right)$$

$$= \frac{1}{\sqrt{2}} \zeta_{np}. \tag{5.96}$$

The secular equation is given by

$$\begin{vmatrix} F_0 - 5F_2 + \frac{1}{2}\zeta_{np} - E & \frac{1}{\sqrt{2}}\zeta_{np} \\ \\ \frac{1}{\sqrt{2}}\zeta_{np} & F_0 + F_2 - E \end{vmatrix} = 0, \tag{5.97}$$

and solving for the energies, we obtain

$$0 = \left(F_0 - 5F_2 + \frac{1}{2}\zeta_{np} - E\right)(F_0 + F_2 - E) - \frac{1}{2}\zeta_{np}^2$$

$$= (E - F_0)^2 + \left(4F_2 - \frac{1}{2}\zeta_{np}\right)(E - F_0) - \left(5F_2 - \frac{1}{2}\zeta_{np}\right)F_2 - \frac{1}{2}\zeta_{np}^2, \tag{5.98}$$

from which two solutions for E are obtained. Again examining the limit of $\zeta_{np} \to 0$, we obtain the term values:

$$E(^3P_2) = F_0 - 2F_2 + \frac{1}{4}\zeta_{np} - \left[9F_2^2 - \frac{3}{2}F_2\zeta_{np} + \frac{9}{16}\zeta_{np}^2\right]^{\frac{1}{2}} \tag{5.99}$$

$$E(^1D_2) = F_0 - 2F_2 + \frac{1}{4}\zeta_{np} + \left[9F_2^2 - \frac{3}{2}F_2\zeta_{np} + \frac{9}{16}\zeta_{np}^2\right]^{\frac{1}{2}}. \tag{5.100}$$

$^3\mathbf{P_1}$

Because there are no interacting states with $J = 1$, we get

$$E(^3P_1) = F_0 - 5F_2 - \frac{1}{2}\zeta_{np}. \tag{5.101}$$

C.

FLUORESCENT LIGHTING UTILIZES THE HG $6s6p$ $^3P_1^0 \to 6s^2$ $^1S_0$ UV TRANSITION AT 253.7 NM TO CAUSE A PHOSPHOR IN THE INSIDE WALLS OF THE FLUORESCENT LAMP TO GLOW "WHITE." THE (STRONGLY ALLOWED) RESONANT TRANSITION OF HG CORRESPONDS TO THE TRANSITION $6s6p$ $^1P_1^0 \to 6s^2$ $^1S_0$ AT 184.9 NM. PRESENT ARGUMENTS TO EXPLAIN WHY THE HG 253.7 NM LINE HAS SIGNIFICANT INTENSITY, ALTHOUGH SINGLET-TRIPLET ELECTRIC DIPOLE TRANSITIONS ARE FORBIDDEN IN THE LS COUPLING LIMIT.

The strength of the "triplet $\to$ singlet" transition can be understood if we consider that for heavy atoms, L and S are not good quantum numbers. Consequently, the "singlet" and "triplet" labels are not exact but instead denote a mixture of singlet and triplet states, which makes the transition allowed. For this atomic transition, the energy levels are derived from strong jj coupling. In this case, the total angular momentum quantum number J can be changed by one unit.

APPLICATION 12
DIRECTIONAL CORRELATIONS;
BREAKUP OF A LONG-LIVED COMPLEX

CONSIDER A REACTIVE SCATTERING PROCESS

$$A + BC \rightarrow AB + C. \tag{5.102}$$

AS A FIRST APPROXIMATION, WE REGARD THE REAGENTS AND PRODUCTS TO BE SPINLESS PARTICLES WITH NO INTERNAL ANGULAR MOMENTUM. LET $\hat{\mathbf{k}}$ AND $\hat{\mathbf{k}}'$ SPECIFY THE INITIAL AND FINAL RELATIVE VELOCITY DIRECTIONS IN THE CENTER-OF-MASS FRAME, AND LET $\hat{\boldsymbol{\ell}}$ AND $\hat{\boldsymbol{\ell}}'$ BE THE CORRESPONDING EXTERNAL ORBITAL ANGULAR MOMENTA WHOSE DIRECTIONS ARE PERPENDICULAR TO $\hat{\mathbf{k}}$ AND $\hat{\mathbf{k}}'$, RESPECTIVELY. WE DENOTE THE TOTAL ANGULAR MOMENTUM OF THE SCATTERING SYSTEM BY $\mathbf{j}$. CONSERVATION OF ANGULAR MOMENTUM IMPLIES THAT

$$\mathbf{j} = \hat{\boldsymbol{\ell}} = \hat{\boldsymbol{\ell}}'. \tag{5.103}$$

THE DIRECTION CORRELATION BETWEEN $\hat{\mathbf{k}}$ AND $\hat{\mathbf{k}}'$, WHICH IS PROPORTIONAL TO THE DIFFERENTIAL CROSS SECTION, DEPENDS ONLY ON THE INCLUDED ANGLE θ, WHERE

$$\cos\theta = \hat{\mathbf{k}} \cdot \hat{\mathbf{k}}'. \tag{5.104}$$

THAT IS, WE MAY WRITE THE DIFFERENTIAL CROSS SECTION AS $I(\theta)$. NEXT WE EXPAND $I(\theta)$ IN A COMPLETE SET OF LEGENDRE POLYNOMIALS,

$$I(\theta) = \frac{1}{4\pi} \sum_{\nu=0}^{\infty} (2\nu + 1) A_\nu P_\nu(\cos\theta), \tag{5.105}$$

WITH THE COEFFICIENTS A_ν GIVEN BY

$$A_\nu = \int_0^{2\pi} \mathrm{d}\phi \int_0^\pi I(\theta) P_\nu(\cos\theta) \sin\theta \, \mathrm{d}\theta$$

$$= \langle P_\nu(\cos\theta) \rangle_{\mathrm{av}}$$

$$= \left\langle P_\nu\left(\hat{\mathbf{k}} \cdot \hat{\mathbf{k}}'\right) \right\rangle_{\mathrm{av}}, \tag{5.106}$$

WHERE $\langle\rangle_{\mathrm{av}}$ DENOTES AN AVERAGE OVER ALL POSSIBLE CONFIGURATIONS OF THE SCATTERING SYSTEM.

WE DEFINE THE z AXIS TO BE ALONG $\hat{\boldsymbol{\ell}}$, WITH THE POLAR ANGLES OF $\hat{\mathbf{k}}$ AND $\hat{\mathbf{k}}'$ DENOTED BY (θ_1, ϕ_1) AND (θ_2, ϕ_2), RESPECTIVELY. WE INTRODUCE THE CONDITION THAT THE COMPLEX IS LONG-LIVED, SO THAT THE ANGLE IN THE xy PLANE CONJUGATE TO $\hat{\boldsymbol{\ell}}$ IS UNIFORMLY DISTRIBUTED. THAT IS, THE COLLISION COMPLEX BREAKS UP WITH A RANDOM VALUE OF $\phi_1 - \phi_2$.

A.

SHOW THAT

$$\left\langle P_\nu\left(\hat{\mathbf{k}} \cdot \hat{\mathbf{k}}'\right) \right\rangle_{\mathrm{av}} = \left\langle P_\nu\left(\hat{\mathbf{k}} \cdot \hat{\boldsymbol{\ell}}\right) P_\nu\left(\hat{\mathbf{k}}' \cdot \hat{\boldsymbol{\ell}}'\right) \right\rangle_{\mathrm{av}} \tag{5.107}$$

AND HENCE

$$A_\nu = [P_\nu(0)]^2 \, \sigma. \tag{5.108}$$

From the spherical harmonic addition theorem, we know that

$$\left\langle P_\nu\left(\hat{\mathbf{k}} \cdot \hat{\mathbf{k}}'\right) \right\rangle_{\text{av}} = \frac{4\pi}{2\nu + 1} \left\langle \sum_q Y_{\nu q}^*(\theta_1, \phi_1) \, Y_{\nu q}(\theta_2, \phi_2) \right\rangle_{\text{av}}, \tag{5.109}$$

which can be rewritten as

$$\left\langle P_\nu\left(\hat{\mathbf{k}} \cdot \hat{\mathbf{k}}'\right) \right\rangle_{\text{av}} = \frac{2}{2\nu + 1} \left\langle \sum_q \Theta_{\nu q}(\theta_1) \, \Theta_{\nu q}(\theta_2) \, e^{-iq(\phi_1 - \phi_2)} \right\rangle_{\text{av}}. \tag{5.110}$$

Here we used the definition

$$Y_{\nu q} = \frac{1}{\sqrt{2\pi}} \, \Theta_{\nu q}(\theta) \, e^{iq\phi} \tag{5.111}$$

and the fact that $\Theta_{\nu q}(\theta)$ is a real function. Because the breakup of the long-lived complex occurs with a random value of $\phi_1 - \phi_2$, all possible values of ϕ in equation 5.106 have equal probability. In addition, we may choose $\phi_1 = 0$ and $\phi_2 = \phi$ without loss of generality. We therefore get from equation 5.110,

$$
\begin{aligned}
\left\langle P_\nu(\hat{\mathbf{k}} \cdot \hat{\mathbf{k}}') \right\rangle_{\text{av}} &= \frac{2}{2\nu + 1} \sum_q \int_0^{2\pi} e^{iq\phi} \mathrm{d}\phi \int_0^\pi I(\theta) \, \Theta_{\nu q}(\theta_1) \, \Theta_{\nu q}(\theta_2) \sin\theta \, \mathrm{d}\theta \\[2mm]
&= \frac{4\pi}{2\nu + 1} \sum_q \delta_{q0} \int_0^\pi I(\theta) \, \Theta_{\nu q}(\theta_1) \, \Theta_{\nu q}(\theta_2) \sin\theta \, \mathrm{d}\theta \\[2mm]
&= \frac{4\pi}{2\nu + 1} \int_0^\pi I(\theta) \, \Theta_{\nu 0}(\theta_1) \, \Theta_{\nu 0}(\theta_2) \sin\theta \, \mathrm{d}\theta \\[2mm]
&= 2\pi \int_0^\pi I(\theta) \, P_\nu(\cos\theta_1) \, P_\nu(\cos\theta_2) \sin\theta \, \mathrm{d}\theta \\[2mm]
&= \left\langle P_\nu\left(\hat{\mathbf{k}} \cdot \hat{\boldsymbol{\ell}}\right) P_\nu\left(\hat{\mathbf{k}}' \cdot \hat{\boldsymbol{\ell}}'\right) \right\rangle_{\text{av}}.
\end{aligned}
\tag{5.112}
$$

From the kinematics of the collision, we know that

$$\hat{\mathbf{k}} \perp \hat{\boldsymbol{\ell}} \qquad \text{and} \qquad \hat{\mathbf{k}}' \perp \hat{\boldsymbol{\ell}}', \tag{5.113}$$

so that

$$\hat{\mathbf{k}} \cdot \hat{\boldsymbol{\ell}} = 0 \tag{5.114}$$

and

$$\hat{\mathbf{k}}' \cdot \hat{\boldsymbol{\ell}}' \quad = \quad 0. \tag{5.115}$$

Substituting equations 5.114 and 5.115 into equation 5.112 and using equation 5.106, we obtain

$$A_\nu \quad = \quad \left\langle P_\nu(\hat{\mathbf{k}} \cdot \hat{\mathbf{k}}') \right\rangle_{\mathrm{av}}$$

$$= \quad 2\pi \int_0^\pi I(\theta) \, [P_\nu(0)]^2 \sin\theta \; \mathrm{d}\theta$$

$$= \quad [P_\nu(0)]^2 \, \sigma. \tag{5.116}$$

Here σ represents the total collision cross section, defined by

$$\sigma = \int_0^\pi \mathrm{d}\phi \int_0^\pi I(\theta) \sin\theta \; \mathrm{d}\theta. \tag{5.117}$$

B.

EXPLAIN WHY THIS RESULT IMPLIES THAT THE PRODUCT ANGULAR DISTRIBUTION HAS FORWARD-BACK-WARD SYMMETRY IN THE CENTER-OF-MASS FRAME, SO THAT $I(\theta)$ IS SYMMETRIC ABOUT $\theta = \frac{\pi}{2}$.

The general formula for the Legendre polynomial is given by[1]

$$P_n(x) = \frac{1}{2^n} \sum_{k=0}^{[n/2]} \frac{(-1)^k \, (2n - 2k)!}{k! \, (n-k)! \, (n-2k)!} \; x^{n-2k}, \tag{5.118}$$

where $[n/2]$ denotes the greatest integer less than or equal to $n/2$. Let us analyze the behavior of the polynomials for even and odd n. For $n = 2m$, equation 5.118 becomes

$$P_{2m}(x) = \frac{1}{2^{2m}} \sum_{k=0}^{m} \frac{(-1)^k \, (4m - 2k)!}{k! \, (2m-k)! \, (2m-2k)!} \; x^{2(m-k)}, \tag{5.119}$$

whereas for $n = 2m + 1$, equation 5.118 becomes

$$P_{2m+1}(x) \quad = \quad \frac{1}{2^{2m+1}} \sum_{k=0}^{m} \frac{(-1)^k \, (4m - 2k + 2)!}{k! \, (2m-k+1)! \, (2m-2k+1)!} \; x^{2(m-k)+1}. \tag{5.120}$$

It follows from equations 5.119 and 5.120 that

$$P_{2m}(0) = \frac{1}{2^{2m}} \frac{(-1)^m \, (2m)!}{(m!)^2} \tag{5.121}$$

and

$$P_{2m+1}(0) = 0. \tag{5.122}$$

---

[1]Gradshteyn I.S., and Ryzhik I.M. *Tables of Integrals, Series, and Products*, 4th ed. New York, NY: Academic Press; 1980:1025.

From equations 5.106 and 5.116 we see that for a long-lived complex only terms with an even value of ν survive in equation 5.105. Considering equation 5.119 for the special case of $x = \cos\theta$, we see that

$$P_{2m}(\cos\theta) \;=\; \frac{1}{2^{2m}} \sum_{k=0}^{m} \frac{(-1)^k \, (4m - 2k)!}{k! \, (2m - k)! \, (2m - 2k)!} \, (\cos\theta)^{2(m-k)}. \tag{5.123}$$

That is, $P_{2m}(\cos\theta)$ is a symmetric function about $\frac{\pi}{2}$. Because $I(\theta)$ is expressed as a sum of Legendre polynomials with even indices, it follows that it is symmetric about $\theta = \frac{\pi}{2}$, which means that the product angular distribution for a long-lived complex has forward-backward symmetry.

C.

SHOW THAT $I(\theta)$ IS PROPORTIONAL TO $1/\sin\theta$, THAT IS, IT PEAKS AT THE POLES AND FANS OUT AT THE EQUATOR. *Hint*: EXPAND $1/\sin\theta$ IN A LEGENDRE SERIES AND COMPARE IT WITH EQUATION Z-5.12.8.

Inserting equations 5.116, 5.121, and 5.123 into equation 5.105 and considering that only the $\nu = 2m$ terms survive, we obtain

$$I(\theta) = \frac{\sigma}{4\pi} \sum_{m=0}^{\infty} (4m + 1) \left[\frac{(-1)^m (2m)!}{2^{2m}(m!)^2} \right]^2 P_{2m}(\cos\theta). \tag{5.124}$$

The Legendre expansion of $\sin\theta^{-1}$ is given by

$$\sin\theta^{-1} = \frac{1}{4\pi} \sum_{\nu=0}^{\infty} (2\nu + 1) B_\nu \, P_\nu(\cos\theta), \tag{5.125}$$

with the expansion coefficients given by[2]

$$
\begin{aligned}
B_{2m} \;&=\; 2\pi \int_{-1}^{1} (1 - x^2)^{-\frac{1}{2}} P_{2m}(x)\mathrm{d}x \\[2mm]
&=\; 2\pi \left[\frac{\Gamma(m + \frac{1}{2})}{m!} \right]^2 \\[2mm]
&=\; \frac{2\pi^2}{2^{4m}} \frac{[(2m)!]^2}{(m!)^4},
\end{aligned}
\tag{5.126}
$$

for $\nu = 2m$ and 0 for $\nu = 2m + 1$. In equation 5.126 we used the identities $\Gamma(\frac{1}{2}) = \sqrt{\pi}$ for $m = 0$ and

$$
\begin{aligned}
\Gamma(m + \frac{1}{2}) \;&=\; \frac{1 \cdot 3 \cdot 5 \cdots (2m - 1)}{2^m} \sqrt{\pi} \\[2mm]
&=\; \frac{\sqrt{\pi}(2m)!}{2^{2m}m!}
\end{aligned}
\tag{5.127}
$$

---

[2] *Ibid.*, p 822.

for $m > 0$. It follows that

$$\frac{1}{\sin\theta} = \frac{\pi}{2} \sum_{m=0}^{\infty} (4m+1) \left[\frac{(-1)^m (2m)!}{2^{2m}(m!)^2} \right]^2 P_{2m}(\cos\theta). \tag{5.128}$$

Comparing equations 5.124 and 5.128, we can rewrite the intensity as

$$I(\theta) = \frac{\sigma}{2\pi^2} \frac{1}{\sin\theta}. \tag{5.129}$$

D.

WE REMOVE THE RESTRICTION THAT THE REACTANTS AND PRODUCTS ARE SPINLESS PARTICLES. THE TOTAL ANGULAR MOMENTUM OF THE COLLISION COMPLEX IS THEN

$$\mathbf{j} = \boldsymbol{\ell} + \mathbf{s} = \boldsymbol{\ell}' + \mathbf{s}', \tag{5.130}$$

WHERE $\mathbf{s} = \mathbf{J}_A + \mathbf{J}_{BC}$ AND $\mathbf{s}' = \mathbf{J}_{AB} + \mathbf{J}_C$ DENOTE THE ENTRANCE AND EXIT CHANNEL SPINS (ANGULAR MOMENTA). IN THE CLASSICAL LIMIT, EACH ANGULAR MOMENTUM VALUE MAY BE ASSOCIATED WITH A DEFINITE DIRECTION IN SPACE, SPECIFIED BY ITS CORRESPONDING UNIT VECTOR. SUCCESSIVE APPLICATIONS OF THE SPHERICAL HARMONIC ADDITION THEOREM ALLOW US TO WRITE $\left\langle P_\nu \left(\hat{\mathbf{k}} \cdot \hat{\mathbf{k}}' \right) \right\rangle_{\text{av}}$ AS

$$\begin{aligned} \left\langle P_\nu \left(\hat{\mathbf{k}} \cdot \hat{\mathbf{k}}' \right) \right\rangle_{\text{av}} &= \left\langle P_\nu \left(\hat{\mathbf{k}} \cdot \hat{\mathbf{j}} \right) P_\nu \left(\hat{\mathbf{j}} \cdot \hat{\mathbf{k}} \right) \right\rangle_{\text{av}} \\[2ex] &= \left\langle P_\nu \left(\hat{\mathbf{k}} \cdot \hat{\boldsymbol{\ell}} \right) P_\nu \left(\hat{\boldsymbol{\ell}} \cdot \hat{\mathbf{j}} \right) P_\nu \left(\hat{\mathbf{j}} \cdot \hat{\boldsymbol{\ell}}' \right) P_\nu \left(\hat{\boldsymbol{\ell}}' \cdot \hat{\mathbf{k}}' \right) \right\rangle_{\text{av}} \\[2ex] &= \left\langle P_\nu \left(\hat{\mathbf{k}} \cdot \hat{\boldsymbol{\ell}} \right) P_\nu \left(\hat{\mathbf{k}}' \cdot \hat{\boldsymbol{\ell}}' \right) P_\nu \left(\hat{\boldsymbol{\ell}} \cdot \hat{\mathbf{j}} \right) P_\nu \left(\hat{\mathbf{j}} \cdot \hat{\boldsymbol{\ell}}' \right) \right\rangle_{\text{av}}, \end{aligned} \tag{5.131}$$

WHERE WE ASSUMED THAT DIRECTIONAL PROPERTIES OF THE REAGENT AND PRODUCT TRAJECTORIES ARE UNCORRELATED EXCEPT BY THE CONSERVATION OF TOTAL ANGULAR MOMENTUM (DEFINITION OF A SEPARABLE LONG-LIVED COLLISION COMPLEX). HERE THE FIRST LINE OF EQUATION 5.131 IS A CONSEQUENCE OF THE RANDOM DISTRIBUTION OF THE DIHEDRAL ANGLE $\phi_{kjk'}$, AND THE SECOND LINE IS A CONSEQUENCE OF THE RANDOM DISTRIBUTION OF $\phi_{k\ell j}$ AND $\phi_{k'\ell'j}$, WHICH RESULTS BECAUSE $\mathbf{s}$ IS ASSUMED RANDOMLY ORIENTED WITH RESPECT TO $\mathbf{k}$ AND $\mathbf{s}'$. NOTE THAT AN UNOBSERVED DIRECTION IS EQUIVALENT TO A RANDOM ONE BECAUSE EACH LEADS TO AN UNWEIGHTED AVERAGE OVER THE DIHEDRAL ANGLE. SHOW ONCE AGAIN THAT ONLY EVEN VALUES OF ν CONTRIBUTE TO $I(\theta)$, THAT IS, THAT $I(\theta)$ HAS FORWARD-BACKWARD SCATTERING SYMMETRY.

Substituting equations 5.114 and 5.115 into the last line of equation 5.131 gives

$$\left\langle P_\nu \left(\hat{\mathbf{k}} \cdot \hat{\mathbf{k}}' \right) \right\rangle_{\text{av}} = \left[P_\nu(0) \right]^2 \left\langle P_\nu \left(\hat{\boldsymbol{\ell}} \cdot \hat{\mathbf{j}} \right) P_\nu \left(\hat{\mathbf{j}} \cdot \hat{\boldsymbol{\ell}}' \right) \right\rangle_{\text{av}}. \tag{5.132}$$

Because we showed in part B that $P_\nu(0)$ vanishes unless ν is even, it follows once again that $I(\theta)$ has forward-backward scattering symmetry.

E.

In a fully quantum mechanical treatment, the Legendre functions are replaced by appropriate 3-j and 6-j symbols, giving rise to

$$I(\theta) = \frac{1}{4\pi} \sum_{\nu} (2\nu + 1)(-1)^{\ell+s+j+\ell'+s'+j} \, P_{\nu}(\cos\theta) \, [(2\ell+1)(2j+1)(2\ell'+1)(2j+1)]^{\frac{1}{2}}$$

$$\times \langle \ell 0, \nu 0 | \ell 0 \rangle \; \langle \ell'0, \nu 0 | \ell'0 \rangle \begin{Bmatrix} \ell & \nu & \ell \\ j & s & j \end{Bmatrix} \begin{Bmatrix} \ell' & \nu & \ell' \\ j & s' & j \end{Bmatrix} . \tag{5.133}$$

An important consequence of the quantization is that the index ν is restricted to a finite number of even values. What conditions can be placed on the maximum value of ν?

Because of the symmetry properties of the 6-j symbols, there are restrictions on the possible values of ν. For these symbols not to vanish, the following triangle conditions must be satisfied: $\Delta(\ell\nu\ell)$, $\Delta(j\nu j)$, $\Delta(j\nu\ell)$, $\Delta(\ell'\nu\ell')$, and $\Delta(j\nu\ell')$. The maximum possible value of ν that satisfies all of these conditions is the smallest number among $2j$, 2ℓ, and $2\ell'$.

F.

Yang has used group theoretic arguments to prove a general theorem about angular correlation between successive photons, namely, $W(\theta_{if})$ is an even power of $\cos\theta_{if}$, and the highest power in this polynomial is smaller than or equal to $2\ell_i$, $2\ell_f$, or $2j$, whichever is the smallest. Show that the semiclassical arguments presented in the preceeding paragraphs lead to the same conclusion as Yang's theorem.

According to the semiclassical arguments presented in the text, the angular correlations between successive photons are proportional to

$$\langle \ell_i 1, \nu 0 | \ell_i - 1 \rangle \tag{5.134}$$

$$\langle \ell_f 1, \nu 0 | \ell_f - 1 \rangle \tag{5.135}$$

and also to

$$\begin{Bmatrix} \ell_i & \nu & \ell_i \\ j & j_i & j \end{Bmatrix} \tag{5.136}$$

$$\begin{Bmatrix} \ell_f & \nu & \ell_f \\ j & j_f & j \end{Bmatrix} . \tag{5.137}$$

Inspection of the Clebsch-Gordan coefficients in expressions 5.134 and 5.135 reveals that ν should be even in order for the coefficients not to vanish. Inspection of the 6-j symbols in expressions 5.136 and 5.137 reveals that the maximum allowed value of ν is the smallest number among $2j$, $2\ell_i$, and $2\ell_f$, which follows from the same argument as in Part E.

APPLICATION 13
ORIENTATION AND ALIGNMENT

A.

PROVE THAT

$$J_{\pm k}^{(k)} = \left[J_{\pm 1}^{(1)}\right]^k \tag{5.138}$$

Hint: CONTRACT $J_{\pm 1}^{(1)}$ AND $J_{\pm 1}^{(1)}$ TO FORM THE SPHERICAL TENSOR $J_{\pm 2}^{(2)}$ AND MAKE AN INDUCTIVE ARGUMENT.

Using equation Z-5.40 to contract two first-rank tensors to yield a second-rank tensor with $k = 2$ and $q = +2$, we obtain

$$\left[J^{(1)} \otimes J^{(1)}\right]_2^{(2)} = \sum_m \langle 1\ m, 1\ 2-m | 2\ 2\rangle\ J_m^{(1)}\ J_{2-m}^{(1)}$$

$$= \langle 1\ 1, 1\ 1 | 2\ 2\rangle\ J_1^{(1)}\ J_1^{(1)}$$

$$= J_1^{(1)}\ J_1^{(1)}. \tag{5.139}$$

Similarly for $k = 2$, $q = -2$, we obtain

$$\left[J^{(1)} \otimes J^{(1)}\right]_{-2}^{(2)} = \sum_m \langle 1\ m, 1\ -2-m | 2\ -2\rangle\ J_m^{(1)}\ J_{-2-m}^{(1)}$$

$$= \langle 1\ -1, 1\ -1 | 2\ -2\rangle\ J_{-1}^{(1)}\ J_{-1}^{(1)}$$

$$= J_{-1}^{(1)}\ J_{-1}^{(1)}. \tag{5.140}$$

It follows that

$$J_{\pm 2}^{(2)} = \left[J_{\pm 1}^{(1)}\right]^2. \tag{5.141}$$

Using mathematical induction, we assume that equation 5.138 is true for k and prove that it also holds for $k + 1$. Using equations Z-5.36 and Z-5.39 to contract tensors of rank k and rank 1, we obtain

$$\left[J^{(k)} \otimes J^{(1)}\right]_{k+1}^{(k+1)} = \sum_m \langle k\ m, 1\ k+1-m | k+1, k+1\rangle\ J_m^{(k)}\ J_{k+1-m}^{(1)}$$

$$= \quad \langle k\ k, 1\ 1 | k+1, k+1 \rangle\ J_k^{(k)}\ J_1^{(1)}$$

$$= \quad J_k^{(k)}\ J_1^{(1)}, \tag{5.142}$$

and

$$\left[J^{(k)} \otimes J^{(1)} \right]_{-k-1}^{(k+1)} = \sum_m \langle k\ m, 1\ \ -k-1-m | k+1, -k-1 \rangle\ J_m^{(k)}\ J_{-k-1-m}^{(1)}$$

$$= \quad \langle k\ -k, 1\ -1 | k+1, -k-1 \rangle\ J_{-k}^{(k)}\ J_{-1}^{(1)}$$

$$= \quad J_{-k}^{(k)}\ J_{-1}^{(1)}. \tag{5.143}$$

In evaluating the sum in equation 5.142, we recognize that the only nonzero Clebsch-Gordan coefficient is that with $m = k$, whereas in equation 5.143 only $m = -k$ survives. Substituting equation 5.138 into equations 5.142 and 5.143 gives the desired result,

$$J_{\pm(k+1)}^{(k+1)} = \left[J_{\pm 1}^{(1)} \right]^{k+1}. \tag{5.144}$$

B.

USE THE RELATIONS

$$\left[J_z, J_q^{(k)} \right] = q J_q^{(k)} \tag{5.145}$$

AND

$$\left[J_\pm, J_q^{(k)} \right] = [k(k+1) - q(q \pm 1)]^{\frac{1}{2}} J_{q \pm 1}^{(k)} \tag{5.146}$$

TO DERIVE $J_{\pm 1}^{(2)}$ AND $J_0^{(2)}$ FROM $J_{\pm 2}^{(2)}$.

Applying equation 5.146 to $J_2^{(2)}$ gives

$$\left[J_-, J_2^{(2)} \right] = 2 J_1^{(2)}. \tag{5.147}$$

But we also know from equations Z-5.15 and 5.141 that

$$J_2^{(2)} = \left[J_1^{(1)} \right]^2 = \frac{1}{2} J_+^2. \tag{5.148}$$

From the properties of ladder operators (equation Z-1.14), we readily show that

$$\left[J_-, J_+^2 \right] = [J_-, J_+] J_+ + J_+\ [J_-, J_+]$$

$$= -2\left(J_z J_+ + J_+ J_z\right)$$

$$= -2\left(J_+ + 2J_+ J_z\right). \tag{5.149}$$

From equations 5.147 through 5.149 we obtain the desired result,

$$J_1^{(2)} = \frac{1}{2}\left[J_-, J_2^{(2)}\right]$$

$$= \frac{1}{4}\left[J_-, J_+^2\right]$$

$$= -\frac{1}{2} J_+ \left(2J_z + 1\right). \tag{5.150}$$

Similarly, applying a raising operator to $J_{-2}^{(2)}$ gives

$$\left[J_+, J_{-2}^{(2)}\right] = 2J_{-1}^{(2)}. \tag{5.151}$$

Recalling equation Z-5.15, we know that

$$J_{-2}^{(2)} = \frac{1}{2} J_-^2. \tag{5.152}$$

It follows that

$$J_{-1}^{(2)} = \frac{1}{2}\left[J_+, \frac{1}{2} J_-^2\right]$$

$$= \frac{1}{2} J_- \left(2J_z - 1\right). \tag{5.153}$$

Finally, we apply a lowering operator to $J_1^{(2)}$ and obtain

$$\left[J_-, J_1^{(2)}\right] = \sqrt{6} J_0^{(2)}. \tag{5.154}$$

From equation 5.150 we know that

$$J_1^{(2)} = -J_+ J_z - \frac{1}{2} J_+. \tag{5.155}$$

Applying a lowering operator to each term on the right side of the above equation and using equations Z-1.14, we get

$$[J_-, J_+ J_z] = -2J_z^2 + J_+ J_- \tag{5.156}$$

and

$$[J_-, J_+] = -2J_z. \tag{5.157}$$

Inserting equations 5.155 through 5.157 into equation 5.154 gives

$$J_0^{(2)} = \frac{1}{\sqrt{6}} \left[2J_z^2 - J_+ J_- + J_z \right].$$ (5.158)

We also know that

$$J^2 = J_z^2 + \frac{1}{2} \left(J_+ J_- + J_- J_+ \right)$$

$$= J_z^2 + J_+ J_- - J_z,$$ (5.159)

or, equivalently,

$$3J_z^2 - J^2 = 2J_z^2 - J_+ J_- + J_z.$$ (5.160)

Inserting equation 5.160 into equation 5.158 gives the desired result,

$$J_0^{(2)} = \frac{3 J_z^2 - J^2}{\sqrt{6}}.$$ (5.161)

C.

THE ANGULAR MOMENTUM TENSORS SATISFY THE RELATION

$$\left(J_q^{(k)} \right)^\dagger = (-1)^q \, J_{-q}^{(k)}.$$ (5.162)

BY EXPLICIT REFERENCE TO EQUATIONS 5.148, 5.150, AND 5.160, SHOW THAT $\left(J_{+2}^{(2)} \right)^\dagger = J_{-2}^{(2)}$, $\left(J_{+1}^{(2)} \right)^\dagger = -J_{-1}^{(2)}$, AND $\left(J_0^{(2)} \right)^\dagger = J_0^{(2)}$. *Hint*: RECALL THAT $(AB)^\dagger = B^\dagger A^\dagger$, WHERE A AND B ARE TWO OPERATORS.

Taking the Hermitian adjoint of equation 5.148 gives

$$\left(J_2^{(2)} \right)^\dagger = \frac{1}{2} \left(J_+^2 \right)^\dagger$$

$$= \frac{1}{2} J_+^\dagger \, J_+^\dagger$$

$$= \frac{1}{2} J_- \, J_-$$

$$= J_{-2}^{(2)}.$$ (5.163)

Repeating this procedure with equation 5.150 and using the properties of ladder operators (equation Z-1.14), we obtain

$$\left(J_{+1}^{(2)} \right)^\dagger = -\frac{1}{2} \left(J_+ \left(2J_z + 1 \right) \right)^\dagger$$

$$= -\frac{1}{2}(2J_z + 1)J_-$$

$$= -\frac{1}{2}J_-(2J_z - 1)$$

$$= -J_{-1}^{(2)}. \qquad (5.164)$$

Finally, taking the Hermitian adjoint of equation 5.161 gives

$$\left(J_0^{(2)}\right)^\dagger = \frac{1}{\sqrt{6}}\left(3J_z^2 - J^2\right)^\dagger$$

$$= \frac{1}{\sqrt{6}}\left(3J_z^2 - J^2\right)$$

$$= J_0^{(2)}. \qquad (5.165)$$

D.

WE CONSIDER A SYSTEM IN A DEFINITE ANGULAR MOMENTUM STATE. IF IT IS IN A PURE STATE, THEN ITS STATE VECTOR ψ CAN BE DESCRIBED BY

$$|\psi\rangle = \sum_M a_M |JM\rangle, \qquad (5.166)$$

WHERE $|JM\rangle$ IS A COMPLETE SET OF EIGENVECTORS. IN THE CASE OF A MIXED STATE, THE SYSTEM CAN BE REPRESENTED BY A DENSITY OPERATOR ρ DEFINED BY

$$\rho = \sum_i W^{(i)} \left|\psi^{(i)}\right\rangle \left\langle\psi^{(i)}\right|. \qquad (5.167)$$

THE ELEMENT OF THE DENSITY MATRIX IS GIVEN BY

$$\rho_{MM'} = \sum_i W^{(i)} a_{M'}^{(i)*} a_M^{(i)}. \qquad (5.168)$$

PROVE THAT $Tr[\rho] = 1$.

$$Tr[\rho] = \sum_M \rho_{MM}$$

$$= \sum_M \sum_i W^{(i)} a_M^{(i)*} a_M^{(i)}$$

$$= \sum_i W^{(i)} \sum_M \left| a^{(i)} \right|^2$$

$$= \sum_i W^{(i)}$$

$$= 1, \tag{5.169}$$

where we have used equations Z-5.13.22 and Z-5.13.25.

E.

PROVE THAT ρ IS HERMITIAN, THAT IS, $\langle J\,M' | \rho | J\,M \rangle^* = \langle J\,M | \rho | J\,M' \rangle$.

$$\rho_{MM'} = \langle J\,M | \rho | J\,M' \rangle$$

$$= \sum_i W^{(i)}\, a_{M'}^{(i)*}\, a_M^{(i)}$$

$$= \left[\sum_i W^{(i)*}\, a_{M'}^{(i)}\, a_M^{(i)*} \right]^*$$

$$= \langle J\,M' | \rho | J\,M \rangle^*$$

$$= \rho_{M'M}^*. \tag{5.170}$$

F.

SHOW THAT FOR A SYSTEM IN A PURE STATE OF DEFINITE ANGULAR MOMENTUM $\mathbf{J}$, THE NUMBER OF REAL PARAMETERS NECESSARY TO SPECIFY THE DENSITY MATRIX IS $4J$.

The system in a pure state of definite angular momentum $\mathbf{J}$ can be described by a state vector $|\Psi\rangle$ as in equation Z-5.13.21

$$|\Psi\rangle = \sum_M a_{JM} |JM\rangle. \tag{5.171}$$

There are $2J + 1$ complex values of a_{JM}, which are specified by $4J + 2$ real parameters. Normalization,

$$\sum_M |a_{JM}|^2 = 1, \tag{5.172}$$

reduces the number of independent parameters by one. Because only the relative phases of a_{JM} matter, the total number of real parameters is reduced by one more, giving a total of $4J$ independent values.

G.

VERIFY EQUATION Z-5.13.35.

We consider the case of $q = 0$ and $M = M' = 0$ because the reduced matrix elements are independent of M:

$$\left\langle J0|J_0^{(2)}|J0 \right\rangle = (-1)^J \begin{pmatrix} J & 2 & J \\ 0 & 0 & 0 \end{pmatrix} \left\langle J||J^{(2)}||J \right\rangle$$

$$= \frac{-2J(J+1)}{[(2J+3)(2J+2)(2J+1)\,2J\,(2J-1)]^{\frac{1}{2}}} \left\langle J||J^{(2)}||J \right\rangle. \tag{5.173}$$

We can also use equation 5.161 to evaluate the left side of equation 5.173:

$$\left\langle J0|J_0^{(2)}|J0 \right\rangle = \frac{1}{\sqrt{6}} \left\langle J0|(3J_z^2 - J^2)|J0 \right\rangle$$

$$= \frac{-1}{\sqrt{6}} J(J+1). \tag{5.174}$$

By comparing equations 5.173 and 5.174, we obtain the desired result,

$$\left\langle J||J^{(2)}||J \right\rangle = \frac{[(2J+3)(J+1)(2J+1)J(2J-1)]^{\frac{1}{2}}}{\sqrt{6}}. \tag{5.175}$$

H.

SHOW THAT AN AXIALLY SYMMETRIC SYSTEM THAT ALSO POSSESSES THE ADDITIONAL SYMMETRY OF A PLANE OF REFLECTION PERPENDICULAR TO THE AXIAL SYMMETRY AXIS HAS ONLY NONVANISHING MULTIPOLE MOMENTS WITH EVEN RANK k, THAT IS, IT SHOWS ONLY ALIGNMENT. IT FOLLOWS THAT PHOTODISSOCIATION OR PHOTOIONIZATION OF AN ISOTROPIC GAS SAMPLE BY A BEAM OF PLANE-POLARIZED OR UNPOLARIZED LIGHT CAN PRODUCE ONLY PHOTOFRAGMENTS THAT ARE ALIGNED BUT NOT ORIENTED. SIMILARLY, BEAM-GAS OR BEAM-BEAM SCATTERING IN WHICH A BEAM OF ISOTROPIC PARTICLES IMPINGES ON AN ISOTROPIC SAMPLE OR ON A BEAM OF ATOMS OR MOLECULES CAN PRODUCE ONLY PRODUCTS THAT ARE ALIGNED (IN THE COORDINATE FRAME IN WHICH THE QUANTIZATION AXIS COINCIDES WITH THE AXIS OF CYLINDRICAL SYMMETRY).

For an axially symmetric system, we know that the only nonvanishing multipole moments are $\left\langle J_0^{(k)} \right\rangle$, which are given by

$$\left\langle J_0^{(k)} \right\rangle = \sum_M \rho_{MM} (-1)^{J-M} \begin{pmatrix} J & k & J \\ -M & 0 & M \end{pmatrix} \left\langle J||J^{(k)}||J \right\rangle. \tag{5.176}$$

If the system possesses additional symmetry of a reflection plane perpendicular to the axial symmetry axis, $\rho_{MM} = \rho_{-M-M}$, and equation 5.176 reduces to

$$\left\langle J_0^{(k)} \right\rangle = \left\langle J||J^{(k)}||J \right\rangle \left\{ \sum_{M>0}^{J} \rho_{MM} \left[(-1)^{J-M} \begin{pmatrix} J & k & J \\ -M & 0 & M \end{pmatrix} + (-1)^{J+M} \begin{pmatrix} J & k & J \\ M & 0 & -M \end{pmatrix} \right] \right.$$

$$\left. + g_J \rho_{00} \begin{pmatrix} J & k & J \\ 0 & 0 & 0 \end{pmatrix} \right\}, \tag{5.177}$$

where $g_J = 1$ if J is an integer and 0 otherwise. From the properties of the 3-j symbols, we know that

$$\begin{pmatrix} J & k & J \\ M & 0 & -M \end{pmatrix} = (-1)^{2J+k} \begin{pmatrix} J & k & J \\ -M & 0 & M \end{pmatrix}. \tag{5.178}$$

Inserting equation 5.178 into equation 5.177, we obtain

$$\left\langle J_0^{(k)} \right\rangle = \left\langle J||J^{(k)}||J \right\rangle \left\{ \sum_{M>0}^{J} \rho_{MM} \begin{pmatrix} J & k & J \\ -M & 0 & M \end{pmatrix} [(-1)^{J-M} + (-1)^{J+M+2J+k}] \right.$$

$$\left. + g_J \rho_{00} \begin{pmatrix} J & k & J \\ 0 & 0 & 0 \end{pmatrix} \right\}$$

$$= \left\langle J||J^{(k)}||J \right\rangle \left\{ \sum_{M>0}^{J} \rho_{MM} (-1)^{J-M} \begin{pmatrix} J & k & J \\ -M & 0 & M \end{pmatrix} [1 + (-1)^{2(J+M)+k}] \right.$$

$$\left. + g_J \rho_{00} \begin{pmatrix} J & k & J \\ 0 & 0 & 0 \end{pmatrix} \right\}$$

$$= \left\langle J||J^{(k)}||J \right\rangle \left\{ \sum_{M>0}^{J} (-1)^{J-M} \rho_{MM} \begin{pmatrix} J & k & J \\ -M & 0 & M \end{pmatrix} [1 + (-1)^{k}] \right.$$

$$\left. + g_J \rho_{00} \begin{pmatrix} J & k & J \\ 0 & 0 & 0 \end{pmatrix} \right\} \tag{5.179}$$

where we have used the fact that $2(J + M)$ is always an even integer. By inspecting equation 5.179, we find that $\left\langle J_0^{(k)} \right\rangle$ becomes zero when k is odd.

I.

CONSIDER A $J = 2$ SYSTEM WITH AXIAL SYMMETRY.   LET THE MAGNETIC SUBLEVEL POPULATIONS $N(J, M)$ BE

$$N(2,2) = \frac{1}{3}$$

$$N(2,1) = \frac{1}{6}$$

$$N(2,0) = 0$$

$$N(2,-1) = \frac{1}{6}$$

$$N(2,-2) = \frac{1}{3}$$

$$(5.180)$$

FIND ALL THE NONVANISHING MULTIPOLE MOMENTS $\left\langle J_q^{(k)} \right\rangle$ OF THIS SYSTEM.

From the definition of the multipole moments in equation Z-5.13.32, we know that the largest possible value of k is $J + J = 4$. Because the population is symmetric in M (i.e., there is a plane of symmetry), the odd multipole moments ($k = 1$ and 3) vanish. Because the system is also axially symmetric, moments with $q \neq 0$ must also vanish. It follows that the only nonvanishing multipole moments are $\left\langle J_0^{(0)} \right\rangle$, $\left\langle J_0^{(2)} \right\rangle$, and $\left\langle J_0^{(4)} \right\rangle$.

Normalization gives $\left\langle J_0^{(0)} \right\rangle = 1$. The higher multipole moments are calculated using equation 5.176 as follows:

$$\left\langle J_0^{(2)} \right\rangle = \sum_{M=-2}^{2} \rho_{MM}(-1)^{2-M} \begin{pmatrix} 2 & 2 & 2 \\ M & -M & 0 \end{pmatrix} \left[\frac{2 \times 3 \times 3 \times 5 \times 7}{6} \right]^{\frac{1}{2}}$$

$$= [3 \times 5 \times 7]^{\frac{1}{2}} \sum_{M=-J}^{J} \rho_{MM}(-1)^{2-M}\, (-1)^{2-M} \frac{2\left(3M^2 - 6\right)}{(7 \times 6 \times 5 \times 4 \times 3)^{\frac{1}{2}}}$$

$$= \sqrt{\frac{1}{6}} \sum_{M=-J}^{J} \rho_{MM}\, \left(3M^2 - 6\right)$$

$$= \sqrt{\frac{1}{6}} \left(\frac{1}{3} \times 6 + \frac{1}{6} \times (-3) + \frac{1}{6} \times (-3) + \frac{1}{3} \times 6 \right)$$

$$= \sqrt{\frac{3}{2}}, \tag{5.181}$$

where the reduced matrix element was evaluated using equation 5.175. Similarly,

$$\left\langle J_0^{(4)} \right\rangle = \sum_{M=-2}^{2} \rho_{MM} (-1)^{2-M} \begin{pmatrix} 2 & 2 & 4 \\ M & -M & 0 \end{pmatrix} \left[\frac{2 \times 3 \times 4 \times 3 \times 5 \times 7 \times 9}{70} \right]^{\frac{1}{2}}, \tag{5.182}$$

using equation Z-5.36 for the reduced matrix element. The result is

$$\left\langle J_0^{(4)} \right\rangle = 36 \times \left(\frac{1}{3} \times \frac{1}{3\sqrt{70}} - \frac{1}{6} \times \frac{4}{3\sqrt{70}} \right)$$

$$= -\frac{4}{\sqrt{70}}. \tag{5.183}$$

J.

USE EQUATION Z-5.13.39 TO FIND THE VALUE OF THE ALIGNMENT TENSOR COMPONENT $A_0^{(2)}$ FOR THE SYSTEM DESCRIBED IN PART I. WHAT IS THE AVERAGE [ROOT-MEAN-SQUARE (RMS)] VALUE OF THE COSINE OF THE ANGLE THAT $\mathbf{J}$ MAKES WITH THE QUANTIZATION AXIS?

The quadrupolar alignment tensor is given by

$$A_0^{(2)} = \frac{\sqrt{6}}{J(J+1)} \, \mathbf{Re} \left\langle \left(J | J_q^{(2)} | J \right) \right\rangle$$

$$= \frac{1}{2}, \tag{5.184}$$

where we have set $J = 2$ and used equation 5.181 for the value of the multipole moment.

The RMS value of any function $f(M)$ is given by

$$\langle f(M) \rangle = \left[\sum_{M=-J}^{J} \rho_{MM} \, f(M)^2 \right]^{\frac{1}{2}}. \tag{5.185}$$

In this problem we have

$$f(M) = \cos \theta = \frac{M}{\sqrt{6}}. \tag{5.186}$$

Inserting equation 5.186 into 5.185 gives

$$\langle \cos \theta \rangle_{RMS} = \left[2 \left(\frac{1}{3} \right) \left(\frac{4}{6} \right) + 2 \left(\frac{1}{6} \right) \left(\frac{1}{6} \right) \right]^{\frac{1}{2}}$$

$$= \sqrt{\frac{1}{2}}, \tag{5.187}$$

which corresponds to the angle

$$\theta = \arccos \langle \cos \theta \rangle_{RMS}$$

$$= \frac{\pi}{4}. \tag{5.188}$$

K.

WE CONSIDER THE ORIENTATION AND ALIGNMENT PRODUCED BY PHOTOIONIZATION. SHOW THAT

$$\sigma(j_1 m_1) = K_0 \sum_{j_e} \sum_{j,j'} [(2j+1)(2j'+1)]^{\frac{1}{2}} \left\langle j' || r^{(1)} || j_0 \right\rangle^* \left\langle j || r^{(1)} || j_0 \right\rangle$$

$$\times \sum_{j_t} (2j_t+1) \left\{ \begin{matrix} j_1 & j_e & j' \\ j_0 & 1 & j_t \end{matrix} \right\} \left\{ \begin{matrix} j_1 & j_e & j \\ j_0 & 1 & j_t \end{matrix} \right\} \left(\begin{matrix} j_1 & 1 & j_t \\ m_1 & -q & q-m_1 \end{matrix} \right)^2, \tag{5.189}$$

FROM WHICH IT FOLLOWS THAT

$$\sigma(j_1 m_1) = \sum_{j_t} \sigma(j_t) \langle j_1 m_1, j_t\ q-m_1 | 1q \rangle^2, \tag{5.190}$$

THAT IS, THE CROSS SECTION $\sigma(j_1 m_1)$ HAS THE FORM OF AN INCOHERENT SUMMATION OVER THE ANGULAR MOMENTUM TRANSFER QUANTUM NUMBER. *Hint*: USE THE WIGNER-ECKART THEOREM AND THE IDENTITY Z-4.15 OR Z-4.16.

We start by defining the different sources of angular momentum. Before photoionization, $\mathbf{j} = \mathbf{j_0} + \mathbf{1}$, where $\mathbf{j_0}$ is the initial angular momentum of the target, and the photon carries one unit of angular momentum. After photoionization, we can write $\mathbf{j} = \mathbf{j_1} + \mathbf{j_e}$, where $\mathbf{j_1}$ is the final angular momentum of the target, and $\mathbf{j_e}$ is the angular momentum of the ejected electron. We define the transferred angular momentum $\mathbf{j_t}$ as $\mathbf{j_t} = \mathbf{j_e} - \mathbf{j_0} = \mathbf{1} - \mathbf{j_1}$. The photoionization cross section is given as

$$\sigma(j_1 m_1) = K_0 \sum_{m_0} \sum_{j_e m_e} \sum_{jm} \sum_{j'm'} \langle j_1 m_1, j_e m_e | j'm' \rangle \langle j_1 m_1, j_e m_e | jm \rangle$$

$$\times \left\langle j'm' \left| r_q^{(1)} \right| j_0 m_0 \right\rangle^* \left\langle jm \left| r_q^{(1)} \right| j_0 m_0 \right\rangle$$

$$= K_0 \sum_{m_0} \sum_{j_e m_e} \sum_{jm} \sum_{j'm'} (-1)^{j_1-j_e+m'} (2j'+1)^{\frac{1}{2}} \left(\begin{matrix} j_1 & j_e & j' \\ m_1 & m_e & -m' \end{matrix} \right)$$

$$\times (-1)^{j_1-j_e+m} (2j+1)^{\frac{1}{2}} \left(\begin{matrix} j_1 & j_e & j \\ m_1 & m_e & -m \end{matrix} \right) (-1)^{j'-m'} \left(\begin{matrix} j' & 1 & j_0 \\ -m' & q & m_0 \end{matrix} \right)$$

$$\times \ \left\langle j'||r^{(1)}||j_0\right\rangle^* \ (-1)^{j-m} \left(\begin{array}{ccc} j & 1 & j_0 \\ -m & q & m_0 \end{array} \right) \ \left\langle j||r^{(1)}||j_0\right\rangle, \tag{5.191}$$

where we have used the Wigner-Eckart theorem to expand the 3-j symbols. Rearranging the terms in the previous expression, we obtain

$$\sigma(j_1 m_1) \ = \ K_0 \sum_{j_e, j, j'} \sqrt{(2j+1)(2j'+1)} \left\langle j'||r^{(1)}||j_0\right\rangle^* \left\langle j||r^{(1)}||j_0\right\rangle$$

$$\times \sum_{m_0, m, m', m_e} \left(\begin{array}{ccc} j_1 & j_e & j' \\ m_1 & m_e & -m' \end{array} \right) \left(\begin{array}{ccc} j_1 & j_e & j \\ m_1 & m_e & -m \end{array} \right)$$

$$\times (-1)^{2j_1 - 2j_e + j + j'} \left(\begin{array}{ccc} j' & 1 & j_0 \\ -m' & q & m_0 \end{array} \right) \left(\begin{array}{ccc} j & 1 & j_0 \\ -m & q & m_0 \end{array} \right).$$

$$\tag{5.192}$$

Using the properties of the 3-j symbols (equations Z-2.30 and Z-2.31), we can reexpress two of them as

$$\left(\begin{array}{ccc} j & 1 & j_0 \\ -m & q & m_0 \end{array} \right) = \left(\begin{array}{ccc} j_0 & 1 & j \\ -m_0 & -q & m \end{array} \right) \tag{5.193}$$

and

$$\left(\begin{array}{ccc} j' & 1 & j_0 \\ -m' & q & m_0 \end{array} \right) = \left(\begin{array}{ccc} j_0 & 1 & j' \\ -m_0 & -q & m' \end{array} \right). \tag{5.194}$$

Using equation Z-4.16, we rewrite the pairs of 3-j symbols in equation 5.192 as

$$\left(\begin{array}{ccc} j_1 & j_e & j \\ m_1 & m_e & -m \end{array} \right) \left(\begin{array}{ccc} j_0 & 1 & j \\ -m_0 & -q & m \end{array} \right)$$

$$= \sum_{j_t, m_t} (2j_t + 1)(-1)^{j_1 + j_e - j + j_0 + 1 + j_t - m_1 + m_0}$$

$$\times \left\{ \begin{array}{ccc} j_1 & j_e & j \\ j_0 & 1 & j_t \end{array} \right\} \left(\begin{array}{ccc} 1 & j_1 & j_t \\ -q & m_1 & m_t \end{array} \right) \left(\begin{array}{ccc} j_e & j_0 & j_t \\ m_e & -m_0 & -m_t \end{array} \right) \tag{5.195}$$

and

$$\left(\begin{array}{ccc} j_1 & j_e & j' \\ m_1 & m_e & -m' \end{array} \right) \left(\begin{array}{ccc} j_0 & 1 & j' \\ -m_0 & -q & m' \end{array} \right)$$

$$= \sum_{j_t', m_t'} (2j_t' + 1)(-1)^{j_1 + j_e - j' + j_0 + 1 + j_t' - m_1 + m_0}$$

$$\times \begin{Bmatrix} j_1 & j_e & j' \\ j_0 & 1 & j'_t \end{Bmatrix} \begin{pmatrix} 1 & j_1 & j'_t \\ -q & m_1 & m'_t \end{pmatrix} \begin{pmatrix} j_e & j_0 & j'_t \\ m_e & -m_0 & -m'_t \end{pmatrix}. \qquad (5.196)$$

Inserting equations 5.195 and 5.196 into 5.192 and collecting powers of (-1), we obtain

$$\sigma(j_1 m_1) = K_0 \sum_{j_e, j, j'} \sqrt{(2j+1)(2j'+1)}(-1)^{2j_0 + j_t + j_{t'}} \left\langle j' \| r^{(1)} \| j_0 \right\rangle^* \left\langle j \| r^{(1)} \| j_0 \right\rangle$$

$$\times \sum_{j_t, m_t} \sum_{j'_t, m'_t} (2j_t + 1)(2j'_t + 1) \begin{Bmatrix} j_1 & j_e & j \\ j_0 & 1 & j_t \end{Bmatrix}$$

$$\times \begin{Bmatrix} j_1 & j_e & j' \\ j_0 & 1 & j'_t \end{Bmatrix} \begin{pmatrix} 1 & j_1 & j_t \\ -q & m_1 & m_t \end{pmatrix} \begin{pmatrix} 1 & j_1 & j'_t \\ -q & m_1 & m'_t \end{pmatrix}$$

$$\times \sum_{m_e, m_0} \begin{pmatrix} j_e & j_0 & j_t \\ m_e & -m_0 & -m_t \end{pmatrix} \begin{pmatrix} j_e & j_0 & j'_t \\ m_e & -m_0 & -m'_t \end{pmatrix}.$$

$$(5.197)$$

Using the orthogonality property of the 3-j symbols,

$$\sum_{m_e, m_0} \begin{pmatrix} j_e & j_0 & j_t \\ m_e & -m_0 & m_t \end{pmatrix} \begin{pmatrix} j_e & j_0 & j'_t \\ m_e & -m_0 & m'_t \end{pmatrix} = (2j_t + 1)^{-1} \delta_{j_t j'_t} \delta_{m_t m'_t}, \qquad (5.198)$$

we may simplify equation 5.197 to obtain the desired result of equation 5.189.
. This result can be further simplified by noting that

$$\begin{pmatrix} j_1 & 1 & j_t \\ m_1 & -q & q - m_1 \end{pmatrix}^2 = \frac{1}{3} \langle j_1 m_1, j_t q - m_1 | 1q \rangle^2, \qquad (5.199)$$

and defining the partial cross section,

$$\sigma(j_t) = \frac{1}{3} K_0 \sum_{j_e, j, j'} \sqrt{(2j+1)(2j'+1)} \left\langle j' \| r^{(1)} \| j_0 \right\rangle^* \left\langle j \| r^{(1)} \| j_0 \right\rangle$$

$$\times \quad (2j_t + 1) \begin{Bmatrix} j_1 & j_e & j' \\ j_0 & 1 & j_t \end{Bmatrix} \begin{Bmatrix} j_1 & j_e & j \\ j_0 & 1 & j_t \end{Bmatrix}. \qquad (5.200)$$

Combining equations 5.189, 5.199, and 5.200, we obtain

$$\sigma(j_1 m_1) = \sum_{j_t} \sigma(j_t) \langle j_1 m_1, j_t q - m_1 | 1q \rangle^2. \qquad (5.201)$$

L.

WE WANT TO CALCULATE THE ORIENTATION,

$$O_0^{(1)}(j_1) = \frac{\sum_{m_1} \sigma(j_1 m_1) m_1 [j_1(j_1+1)]^{-\frac{1}{2}}}{\sum_{m_1} \sigma(j_1 m_1)}, \tag{5.202}$$

AND ALIGNMENT,

$$A_0^{(2)}(j_1) = \frac{\sum_{m_1} \sigma(j_1 m_1) \left[3m_1^2 - j_1(j_1+1)\right] / [j_1(j_1+1)]}{\sum_{m_1} \sigma(j_1 m_1)}, \tag{5.203}$$

AND FIND THE CONTRIBUTION FOR EACH ANGULAR MOMENTUM TRANSFER CONTINUUM CHANNEL. FIND EXPRESSIONS FOR $O_0^{(1)}(j_1, j_t; q)$ AND $A_0^{(2)}(j_1, j_t; q)$, AND SHOW THAT

$$O_0^{(1)}(j_1, j_t; q) \quad = -\tfrac{1}{2} q \left[j_1(j_1+1)\right]^{\frac{1}{2}} \qquad\qquad j_t = j_1 + 1$$

$$O_0^{(1)}(j_1, j_t; q) \quad = \tfrac{1}{2} q \left[j_1(j_1+1)\right]^{-\frac{1}{2}} \qquad\qquad j_t = j_1$$

$$O_0^{(1)}(j_1, j_t; q) \quad = \tfrac{1}{2} q \left[(j_1+1)/j_1\right]^{\frac{1}{2}} \qquad\qquad j_t = j_1 - 1$$

AND THAT

$$A_0^{(2)}(j_1, j_t; q = 0) \quad = -\tfrac{2}{5} + \tfrac{3}{5(j_1+1)} \qquad\qquad j_t = j_1 + 1$$

$$A_0^{(2)}(j_1, j_t; q = 0) \quad = \tfrac{4}{5} - \tfrac{3}{5 j_1(j_1+1)} \qquad\qquad j_t = j_1$$

$$A_0^{(2)}(j_1, j_t; q = 0) \quad = -\tfrac{2}{5} + \tfrac{3}{5 j_1} \qquad\qquad j_t = j_1 - 1.$$

Hint: USE THE RESULTS OF EQUATIONS Z-1.97 TO Z-1.99 OF PROBLEM SET 1.

Summing equation 5.201 over m_1, we obtain

$$\sum_{m_1} \sigma(j_1 m_1) \quad = \quad \sum_{j_t} \sigma(j_t) \sum_{m_1} \langle j_1 m_1, j_t \; q - m_1 | 1q \rangle^2$$

$$= \quad \sum_{j_t} \sigma(j_t). \tag{5.204}$$

Substituting equations 5.201 and 5.204 into equation 5.202, we obtain

$$O_0^{(1)}(j_1) \quad = \quad [j_1(j_1+1)]^{-\frac{1}{2}} \frac{\sum_{j_t} \sigma(j_t) \sum_{m_1} m_1 \langle j_1 m_1, j_t \; q - m_1 | 1q \rangle^2}{\sum_{j_t} \sigma(j_t)}. \tag{5.205}$$

Defining the quantity

$$O_0^{(1)}(j_1, j_t; q) \quad = \quad [j_1(j_1+1)]^{-\frac{1}{2}} \sum_{m_1} m_1 \langle j_1 m_1, j_t \; q - m_1 | 1q \rangle^2, \tag{5.206}$$

we can rewrite equation 5.205 as

$$O_0^{(1)}(j_1) = \frac{\sum_{j_t} \sigma(j_t) O_0^{(1)}(j_1, j_t; q)}{\sum_{j_t} \sigma(j_t)}. \tag{5.207}$$

We can use equation 5.206 to calculate $O_0^{(1)}(j_1, j_t; q)$ for different values of q and j_t. We note that q can have values of $-1, 0,$ and $+1$, and we find that j_t is restricted to $j_1 + 1, j_1,$ and $j_1 - 1$.

Consider first the case of $j_t = j_1 + 1$.

$$O_0^{(1)}(j_1, j_t; q) = \sum_{m_1} \langle j_1 m_1, j_1 + 1 \; q - m_1 | 1 q \rangle^2 \frac{m_1}{[j_1(j_1+1)]^{\frac{1}{2}}}$$

$$= \sum_{m_1} \langle j_1 m_1, 1 \; -q | j_1 + 1 \; m_1 - q \rangle^2 \frac{3}{(2j_1+3)} \frac{m_1}{[j_1(j_1+1)]^{\frac{1}{2}}}, \tag{5.208}$$

where we have used equation Z-2.26 to rearrange the order of the arguments in the Clebsch-Gordan coefficient.

For $q = -1$,

$$O_0^{(1)}(j_1, j_t; q) = \sum_{m_1} \frac{(j_1 + m_1 + 1)(j_1 + m_1 + 2)}{(2j_1+1)(2j_1+2)} \frac{3}{(2j_1+3)} \frac{m_1}{[j_1(j_1+1)]^{\frac{1}{2}}}$$

$$= \frac{3}{(2j_1+1)(2j_1+2)(2j_1+3)[j_1(j_1+1)]^{\frac{1}{2}}}$$

$$\times \sum_{m_1} \left[m_1^3 + (2j_1+3)m_1^2 + (j_1+1)(j_1+2)m_1 \right], \tag{5.209}$$

where we have used Table Z-2.4 to evaluate the Clebsch-Gordan coefficient. The first and third terms in the sum over m_1 vanish because m_1^3 and m_1 are odd functions. The sum over the middle term can be evaluated using equation Z-1.98, which gives

$$\sum_{m_1=-j_1}^{j_1} m_1^2 = \frac{1}{3} j_1(j_1+1)(2j_1+1). \tag{5.210}$$

Inserting equation 5.210 into equation 5.209, we obtain

$$O_0^{(1)}(j_1, j_t; q) = \frac{3}{(2j_1+1)(2j_1+2)(2j_1+3)[j_1(j_1+1)]^{\frac{1}{2}}} \frac{j_1(j_1+1)(2j_1+1)(2j_1+3)}{3}$$

$$= \frac{1}{2} \left[\frac{j_1}{j_1+1} \right]^{\frac{1}{2}}. \tag{5.211}$$

For $q = 0$,

$$O_0^{(1)}(j_1, j_t; q) = \frac{3 \sum_{m_1} m_1(j_1 - m_1 + 1)(j_1 + m_1 + 1)}{(2j_1+1)(j_1+1)[j_1(j_1+1)]^{\frac{1}{2}}(2j_1+3)}$$

$$= \frac{3\sum_{m_1}(j_1+1)^2 m_1 - \sum_{m_1} m_1^3}{(2j_1+1)(j_1+1)[j_1(j_1+1)]^{\frac{1}{2}}(2j_1+3)}$$

$$= 0, \tag{5.212}$$

where we again used the fact that sums over odd powers of m_1 are zero.

For $q = 1$,

$$O_0^{(1)}(j_1, j_t; q) = \frac{3\sum_{m_1} m_1(j_1 - m_1 + 1)(j_1 - m_1 + 2)}{(2j_1+1)(j_1+1)[j_1(j_1+1)]^{\frac{1}{2}}(2j_1+3)}$$

$$= \frac{-3\sum_{m_1}(2j_1+3)m_1^2}{(2j_1+1)(j_1+1)[j_1(j_1+1)]^{\frac{1}{2}}(2j_1+3)}$$

$$= -\frac{1}{2}\left[\frac{j_1}{j_1+1}\right]^{\frac{1}{2}}. \tag{5.213}$$

Therefore, we can say that for $j_t = j_1 + 1$,

$$O_0^{(1)}(j_1, j_t; q) = -\frac{1}{2}\, q\, \left[\frac{j_1}{j_1+1}\right]^{\frac{1}{2}}. \tag{5.214}$$

We now evaluate the case of $j_t = j_1$:

$$O_0^{(1)}(j_1, j_t; q) = \sum_{m_1} \langle j_1 m_1, 1\ -q| j_1\ m_1 - q\rangle^2 \frac{3}{(2j_1+1)} \frac{m_1}{[j_1(j_1+1)]^{\frac{1}{2}}}. \tag{5.215}$$

For $q = -1$

$$O_0^{(1)}(j_1, j_t; q) = \frac{3\sum_{m_1} m_1(j_1 + m_1 + 1)(j_1 - m_1)}{(2j_1+1)[j_1(j_1+1)]^{\frac{1}{2}}2j_1(j_1+1)}$$

$$= -\frac{1}{2}\frac{1}{[j_1(j_1+1)]^{\frac{1}{2}}}. \tag{5.216}$$

Similar calculations for $q = 0$ and $q = +1$ reveal

$$O_0^{(1)}(j_1, j_t; q) = \frac{1}{2}\, q\, \frac{1}{[j_1(j_1+1)]^{\frac{1}{2}}}. \tag{5.217}$$

A similar derivation can be used to evaluate $O_0^{(1)}(j_1, j_1 - 1; q)$.

We turn now to the alignment parameter $A_0^{(2)}(j_1)$. From the definition in equation 5.203, we can write

$$A_0^{(2)}(j_1) = \frac{\sum_{j_t} \sigma(j_t) \sum_{m_1} [3m_1^2 - j_1(j_1+1)] \langle j_1 m_1, j_t\ q - m_1|1q\rangle^2}{j_1(j_1+1) \sum_{j_t} \sigma(j_t)}. \tag{5.218}$$

Defining the quantities,

$$A^{(2)}(j_1, j_t; q) = \frac{\sum_{m_1}[3m_1^2 - j_1(j_1 + 1)]\langle j_1 m_1, j_t\ q - m_1|1q\rangle^2}{j_1(j_1 + 1)}$$

$$= -1 + \frac{3}{j_1(j_1 + 1)}\sum_{m_1} m_1^2 \langle j_1 m_1, j_t\ q - m_1|1q\rangle^2, \tag{5.219}$$

we obtain

$$A_0^{(2)}(j_1) = \frac{\sum_{j_t} \sigma(j_t)\ A^{(2)}(j_1, j_t; q)}{\sum_{j_t} \sigma(j_t)}. \tag{5.220}$$

We consider the case of linearly polarized light ($q = 0$), with $j_t = j_1 + 1$. From equation 5.219,

$$A_0^{(2)}(j_1) = -1 + \frac{3}{j_1(j_1 + 1)}\sum_{m_1} m_1^2 \langle j_1 m_1, j_1 + 1\ - m_1|10\rangle^2$$

$$= -1 + \frac{3}{j_1(j_1 + 1)}\sum_{m_1} m_1^2 \frac{3}{2j_1 + 3}\langle j_1 m_1, 10|j_1 + 1\ m_1\rangle^2$$

$$= -1 + \frac{3}{j_1(j_1 + 1)}\sum_{m_1} m_1^2 \frac{3}{2j_1 + 3}\frac{(j_1 - m_1 + 1)(j_1 + m_1 + 1)}{(2j_1 + 1)(j_1 + 1)}. \tag{5.221}$$

The sum over m_1 is evaluated as

$$\sum_{m_1} m_1^2(j_1 - m_1 + 1)(j_1 + m_1 + 1) = (j_1 + 1)^2\sum_{m_1} m_1^2 - \sum_{m_1} m_1^4$$

$$= \frac{1}{3}(j_1 + 1)^2 j_1(j_1 + 1)(2j_1 + 1)$$

$$- \frac{1}{15}j_1(j_1 + 1)(2j_1 + 1)[3j_1(j_1 + 1) - 1], \tag{5.222}$$

where we have used equations Z-1.98 and Z-1.99. Combining equations 5.221 and 5.222, we obtain

$$A_0^{(2)}(j_1) = -1 + \frac{9}{j_1(j_1 + 1)}\frac{1}{(2j_1 + 1)(j_1 + 1)(2j_1 + 3)}$$

$$\times \left[\frac{1}{3}(j_1 + 1)^3 j_1(2j_1 + 1) - \frac{1}{15}j_1(j_1 + 1)(2j_1 + 1)(3j_1(j_1 + 1) - 1)\right]$$

$$= -1 + \frac{9}{(2j_1 + 1)(2j_1 + 3)}\left[\frac{1}{3}(j_1 + 1)^2 - \frac{1}{15}(3j_1(j_1 + 1) - 1)\right]$$

$$= -\frac{2}{5} + \frac{3}{5(j_1 + 1)}. \tag{5.223}$$

The cases of $j_t = j_1$ and $j_t = j_1 - 1$ are similarly derived.

M.

CONSIDER A ROTATION $\mathbf{R}$ ACTING ON $T(J)_{KQ}$. SHOW THAT

$$\mathbf{R}T(J)_{KQ}\mathbf{R}^{-1} = \sum_{Q'} D_{QQ'}^K(R)T(J)_{KQ'}, \tag{5.224}$$

FROM WHICH IT FOLLOWS THAT $T(J)_{KQ}$ IS A SPHERICAL TENSOR OPERATOR OF RANK K AND COMPONENT Q. THE $T(J)_{KQ}$ ARE CALLED *state multipoles*.

The spherical tensor is defined in equation Z-5.13.54 by the expression

$$T(J)_{KQ} = \sum_{M',M} (-1)^{J-M} \langle JM', J-M|KQ \rangle |JM'\rangle \langle JM|. \tag{5.225}$$

Transforming both sides of this equation, we obtain

$$\mathbf{R}T(J)_{KQ}\mathbf{R}^{-1} = \sum_{M',M} (-1)^{J-M} \langle JM', J-M|KQ \rangle \mathbf{R}|JM'\rangle \langle JM|\mathbf{R}^{-1}$$

$$= \sum_{M',M} (-1)^{J-M} \langle JM', J-M|KQ \rangle \sum_{m'} D_{m'M'}^J |Jm'\rangle \sum_{m} \langle Jm| D_{mM}^{J*}$$

$$= \sum_{M,M'} \sum_{m,m'} (-1)^{J-M}(-1)^{M-m} \langle JM', J-M|KQ \rangle |Jm'\rangle D_{m'M'}^J D_{-m-M}^J \langle Jm|$$

$$= \sum_{m,m'} (-1)^{J-m} |Jm'\rangle \langle Jm| \sum_{M',M} \langle JM', J-M|KQ \rangle D_{m'M'}^J D_{-m-M}^J. \tag{5.226}$$

Substituting a Clebsch-Gordan series (equation Z-3.105) in the preceding result gives

$$\mathbf{R}T(J)_{KQ}\mathbf{R}^{-1} = \sum_{m,m'} (-1)^{J-m} |Jm'\rangle \langle Jm| \sum_{k,Q',q'} \langle Jm', J-m|kQ' \rangle$$

$$\times \sum_{M',M} \langle JM', J-M|kq' \rangle \langle JM', J-M|KQ \rangle D_{Q'q'}^k$$

$$= \sum_{k,Q',q'} \sum_{m,m'} (-1)^{J-m} |Jm'\rangle \langle Jm| \langle Jm', J-m|kQ' \rangle \delta_{kK} \delta_{q'Q} D_{Q'q'}^k$$

$$= \sum_{Q'} \sum_{m,m'} (-1)^{J-m} \langle Jm', J - m | KQ' \rangle |Jm'\rangle \langle Jm| D^K_{Q'Q}$$

$$= \sum_{Q'} D^K_{Q'Q} T(J)_{KQ'}, \tag{5.227}$$

where in the last step we invoked the definition of a spherical tensor operator given in equation 5.225.

N.

INVERT EQUATION Z-5.13.59 TO SHOW THAT

$$\rho_{KQ} = \mathbf{Tr}\left[T^\dagger_{KQ} \, \rho \right] = \left\langle T^\dagger_{KQ} \right\rangle. \tag{5.228}$$

Hint: USE THE ORTHONORMALITY PROPERTIES OF THE CLEBSCH-GORDAN COEFFICIENTS TO PROVE THAT

$$\mathbf{Tr}\left[T^\dagger_{K'Q'} T_{KQ} \, \rho \right] = \delta_{K'K} \, \delta_{Q'Q}, \tag{5.229}$$

WHERE $\left\langle JM \left| T(J)^\dagger_{KQ} \right| JM' \right\rangle = \langle JM' | T(J)_{KQ} | JM \rangle^*.$

We start by expanding the density matrix as a sum of state multipoles,

$$\rho = \sum_{K'Q'} \rho_{K'Q'} \, T_{K'Q'}. \tag{5.230}$$

Next, we multiply equation 5.230 by $T^\dagger_{KQ}$ and take the trace,

$$\mathbf{Tr}\left[T^\dagger_{KQ} \, \rho \right] = \sum_{K'Q'} \rho_{K'Q'} \, \mathbf{Tr}\left[T^\dagger_{KQ} T_{K'Q'} \right]. \tag{5.231}$$

From equation Z-5.13.54, we know that

$$T(J)^\dagger_{KQ} T(J)_{K'Q'} = \sum_{MM'} (-1)^{J-M} \langle JM', J - M | KQ \rangle |JM\rangle \langle JM'|$$

$$\times \sum_{mm'} (-1)^{J-m} \langle Jm', J - m | K'Q' \rangle |Jm'\rangle \langle Jm|$$

$$= \sum_{MM'} \sum_{mm'} (-1)^{2J-M-m} \langle JM', J - M | KQ \rangle$$

$$\times \langle Jm', J - m | K'Q' \rangle |JM\rangle \langle Jm| \delta_{M'm'}$$

$$= \sum_{MM'} \sum_{m} (-1)^{2J-M-m} \langle JM', J - M | KQ \rangle$$

$$\times \langle JM', J \ -m|K'Q'\rangle |JM\rangle \langle Jm| \,. \tag{5.232}$$

Taking the trace of equation 5.232, we get

$$\mathbf{Tr}\left[T(J)_{KQ}^{\dagger} T(J)_{K'Q'}\right] \;=\; \sum_{MM'm} (-1)^{2J-M-m} \langle JM', J \ -M|KQ\rangle \langle JM', J \ -m|K'Q'\rangle \, \delta_{Mm}$$

$$=\; \sum_{MM'} (-1)^{2J-2M} \langle JM', J \ -M|KQ\rangle \langle JM', J \ -M|K'Q'\rangle$$

$$=\; \delta_{KK'}\, \delta_{QQ'} \,, \tag{5.233}$$

where $2J - 2M$ is an even integer. Inserting equation 5.233 into equation 5.231 gives

$$\rho_{KQ} = \mathbf{Tr}\left[T_{KQ}^{\dagger}\, \rho\right] . \tag{5.234}$$

O.

Consider an electronically excited atom or molecule, with initial angular momentum quantum number J_i that makes a spontaneous dipole transition to a final state with angular momentum quantum number J_f. Let the M_i population of the initial state be distributed according to the probability law

$$P(J_i, M_i) \;=\; \frac{1}{2J_i + 1}\left\{1 + 3O_0^{(1)}(J_i)\frac{M_i}{[J_i(J_i+1)]^{\frac{1}{2}}}\right.$$

$$\left. + A_0^{(2)}(J_i)\left[\frac{5J_i(2J_i+1)}{(2J_i+3)(2J_i-1)}\right]\left[\frac{3M_i^2 - J_i(J_i+1)}{J_i(J_i+1)}\right]\right\} . \tag{5.235}$$

That is, the initial state has cylindrical symmetry with an orientation characterized by the value $O_0^{(1)}(J_i)$ and an alignment characterized by the value $A_0^{(2)}(J_i)$. Present arguments to indicate that the probability $P(J_f, M_f)$ of populating the final state J_f, M_f by the radiative transition is

$$P(J_f, M_f) \propto \sum_{M_i} P(J_i, M_i) \langle J_i M_i, 1 \ M_f - M_i|J_f M_f\rangle^2 \,. \tag{5.236}$$

The probability that a system in the initial state $|J_i M_i\rangle$ makes a spontaneous electric dipole transition to a final state $|J_f M_f\rangle$ is given by $P(|J_f M_f\rangle \leftarrow |J_i M_i\rangle) \propto |\langle J_f M_f|\boldsymbol{\mu}\cdot\mathbf{E}|J_i M_i\rangle|^2$, where $\boldsymbol{\mu}$ is the electric dipole operator and $\mathbf{E}$ is the electric field vector of the light. Because both $\boldsymbol{\mu}$ and $\mathbf{E}$ are first-rank tensor (vector) operators we get

$$P(|J_f M_f\rangle \leftarrow |J_i M_i\rangle) \;\propto\; \left|\sum_q \langle J_f M_f|(-1)^q \boldsymbol{\mu}(1,q)\mathbf{E}(1,-q)|J_i M_i\rangle\right|^2$$

$$= \left| \sum_q (-1)^q \left\langle J_f M_f \left| \boldsymbol{\mu}(1,q) \right| J_i M_i \right\rangle \mathbf{E}(1,-q) \right|^2. \tag{5.237}$$

Applying the Wigner-Eckart theorm to the right side of equation 5.237, we obtain

$$P(|J_f M_f\rangle \leftarrow |J_i M_i\rangle) \propto \left| \sum_q (-1)^q (-1)^{J_f - M_f} \begin{pmatrix} J_f & 1 & J_i \\ -M_f & q & M_i \end{pmatrix} \left\langle J_f \| \boldsymbol{\mu}^1 \| J_i \right\rangle \mathbf{E}(1,-q) \right|^2. \tag{5.238}$$

We know from the properties of 3-j symbols that the 3-j symbol on the right side of equation 5.238 vanishes unless $q = M_f - M_i$. Hence we obtain

$$P(|J_f M_f\rangle \leftarrow |J_i M_i\rangle) \quad \propto \quad \begin{pmatrix} J_f & 1 & J_i \\ -M_f & M_f - M_i & M_i \end{pmatrix}^2 = \begin{pmatrix} J_i & 1 & J_f \\ M_i & M_f - M_i & -M_f \end{pmatrix}^2$$

$$\propto \quad \left\langle J_i M_i, 1 \; M_f - M_i \middle| J_f M_f \right\rangle^2. \tag{5.239}$$

The probablility $P(J_f, M_f)$ of populating the final state $|J_f M_f\rangle$ can then be found by

$$P(J_f, M_f) \quad \propto \quad \sum_{M_i} P(J_i, M_i) P(|J_f M_f\rangle \leftarrow |J_i M_i\rangle)$$

$$\propto \quad \sum_{M_i} P(J_i, M_i) \left\langle J_i M_i, 1 \; M_f - M_i \middle| J_f M_f \right\rangle^2. \tag{5.240}$$

APPLICATION 14
NUCLEAR QUADRUPOLE INTERACTIONS

A.

EVEN THESE ELEMENTARY CONSIDERATIONS SUFFICE TO SHOW THAT THE COUPLING OF **J** WITH **I**(CL) IS DOMINANT IN CLCN. DISCUSS HOW THE APPEARANCE OF THE $J'' = 0 \rightarrow J' = 1$ TRANSITION OF CLCN WOULD DIFFER FROM THAT IN FIGURE Z-5.14.1 IF THE COUPLING OF **J** WITH THE NITROGEN NUCLEAR SPIN DOMINATED. WOULD THE TOTAL NUMBER OF LINES CHANGE?

In this case, we denote the nuclear spin of N by $\mathbf{I_1}$ and that of Cl by $\mathbf{I_2}$. Assuming that the coupling of **J** with $\mathbf{I_1}$ dominates that of **J** with $\mathbf{I_2}$, the coupling scheme can be written as

$$\mathbf{J} + \mathbf{I_1} = \mathbf{F_1} \quad \text{and} \quad \mathbf{F_1} + \mathbf{I_2} = \mathbf{F}. \tag{5.241}$$

Again, the $J = 0$ rotational level is not split. For J=1 and $I_1 = 1$, F_1 can be either $0, 1$, or 2. Coupling $F_1 = 0$ with $I_2 = \frac{3}{2}$ gives $F = \frac{3}{2}$; coupling $F_1 = 1$ with $I_2 = \frac{3}{2}$ gives $F = \frac{5}{2}$, $\frac{3}{2}$, or $\frac{1}{2}$; and finally coupling $F_1 = 2$ with $I_2 = \frac{3}{2}$ gives $F = \frac{7}{2}$, $\frac{5}{2}$, $\frac{3}{2}$, or $\frac{1}{2}$. At normal pressures, we expect only the coupling with N to be resolved, producing three lines in the spectrum. At low pressure, the coupling with Cl may be resolved, and the spectrum would contain a singlet, a triplet, and a quartet, for a total of eight lines.

B.

THE ELECTRIC FIELD GRADIENT COUPLING CONSTANT IS DEFINED BY

$$q_{ZZ} = 2 \left\langle \alpha J J | V_0^{(2)} | \alpha J J \right\rangle, \tag{5.242}$$

WHERE $V_q^{(2)}$ ARE COMPONENTS OF THE ELECTRIC FIELD GRADIENT TENSOR AND $|\alpha J J\rangle$ DENOTES THE RIGID ROTOR WAVE FUNCTION, WITH **J** MAKING A MAXIMUM PROJECTION ON THE SPACE-FIXED Z AXIS OF $M_J = J$. SHOW THAT q_{ZZ} VANISHES FOR $J = 0$ OR $J = \frac{1}{2}$, AND SHOW THAT FOR $J \geq 1$,

$$\left\langle \alpha J || V^{(2)} || \alpha J \right\rangle = \frac{1}{2} q_{ZZ} \left[\frac{(2J+1)(2J+2)(2J+3)}{2J(2J-1)} \right]^{\frac{1}{2}}. \tag{5.243}$$

By applying the Wigner-Eckart theorem, we obtain from equation 5.242

$$q_{ZZ} = 2 \begin{pmatrix} J & 2 & J \\ -J & 0 & J \end{pmatrix} \left\langle \alpha J || V^{(2)} || \alpha J \right\rangle. \tag{5.244}$$

It is clear from inspection of the 3-j symbol in equation 5.244 that q_{ZZ} vanishes for $J < 1$ (i.e., for $J = 0$ or $J = \frac{1}{2}$).

For $J \geq 1$, equation 5.244 can be rearranged to give

$$\left\langle \alpha J || V^{(2)} || \alpha J \right\rangle = \frac{1}{2} \left[\begin{pmatrix} J & 2 & J \\ -J & 0 & J \end{pmatrix} \right]^{-1} q_{ZZ}$$

$$= \frac{1}{2} q_{ZZ} \left[\frac{(2J+1)(2J+2)(2J+3)}{2J(2J-1)} \right]^{\frac{1}{2}}. \tag{5.245}$$

C.

CONSIDER THE COUPLED STATES $|JIFM_F\rangle$. FIND AN EXPRESSION FOR THE QUADRUPOLE INTERACTION ENERGY USING FIRST-ORDER PERTURBATION THEORY. *Hint*: SHOW THAT

$$\Delta E = \langle \alpha JIFM_F | \mathbf{V}^{(2)} \cdot \mathbf{Q}^{(2)} |\alpha JIFM_F\rangle$$

$$= \frac{1}{2} \, eq_{zz}Q \, \frac{\frac{3}{4}C(C+1) - I(I+1)J(J+1)}{I(2I-1)J(2J-1)}, \tag{5.246}$$

WHERE

$$C = F(F+1) - J(J+1) - I(I+1). \tag{5.247}$$

THE FACTOR $eq_{zz}Q$ IS CALLED THE *quadrupole coupling* CONSTANT.

In first-order perturbation theory, the energy correction is given by the diagonal matrix element of the perturbing Hamiltonian. For the quadrupole interaction, this matrix element is

$$\Delta E = \langle \alpha JIFM_F | H_Q |\alpha JIFM_F\rangle$$

$$= \langle \alpha JIFM_F | \mathbf{V}^{(2)} \cdot \mathbf{Q}^{(2)} |\alpha JIFM_F\rangle. \tag{5.248}$$

Using equation Z-5.71, we obtain

$$\Delta E = (-1)^{J+I+F} \left\{ \begin{matrix} J & I & F \\ I & J & 2 \end{matrix} \right\} \left\langle \alpha J||\mathbf{V}^{(2)}||\alpha J\right\rangle \left\langle \alpha J||\mathbf{Q}^{(2)}||\alpha J\right\rangle. \tag{5.249}$$

Using Table Z-4.1 to evaluate the 6-j symbol, we obtain

$$\left\{ \begin{matrix} J & I & F \\ I & J & 2 \end{matrix} \right\} = \left\{ \begin{matrix} F & I & J \\ 2 & J & I \end{matrix} \right\}$$

$$= (-1)^{J+I+F} \times 2 \left[(2I+3)(2I+2)(2I+1)2I(2I-1)\right]^{-\frac{1}{2}}$$

$$\times \left[(2J+3)(2J+2)(2J+1)2J(2J-1)\right]^{-\frac{1}{2}} \left[3C(C+1) - 4J(J+1)I(I+1)\right], \tag{5.250}$$

where C is defined in equation 5.247. Inserting equations 5.245, 5.250, Z-5.14.13, and Z-5.14.16 into equation 5.249, we obtain

$$\Delta E = 2 \left[(2I+3)(2I+2)(2I+1)2I(2I-1)\right]^{-\frac{1}{2}} \frac{1}{2}eQ \left[\frac{(2I+1)(2I+2)(2I+3)}{2I(2I-1)}\right]^{\frac{1}{2}}$$

$$\times \left[(2J+3)(2J+2)(2J+1)2J(2J-1)\right]^{-\frac{1}{2}} \frac{1}{2}q_{zz} \left[\frac{(2J+1)(2J+2)(2J+3)}{2J(2J-1)}\right]^{\frac{1}{2}}$$

$$\times \left[3C(C+1)-4J(J+1)I(I+1)\right]$$

$$= \frac{1}{2}eq_{zz}Q \left[\frac{\frac{3}{4}C(C+1)-I(I+1)J(J+1)}{I(2I-1)J(2J-1)}\right]. \qquad (5.251)$$

D.

ESTIMATE THE VALUE OF eqQ FOR ^{35}CL IN ^{35}CLCN FROM THE DATA PROVIDED. SHOW THAT THIS IS APPROXIMATELY -83 MHZ. *Hint*: USE EQUATION 5.251 TO EVALUATE THE MATRIX ELEMENTS $\langle FM_F|H_Q|FM_F\rangle$, AND SHOW THAT $\langle \frac{1}{2}M_F|H_Q|\frac{1}{2}M_F\rangle = -eqQ/4$, $\langle \frac{3}{2}M_F|H_Q|\frac{3}{2}M_F\rangle = eqQ/5$, AND $\langle \frac{5}{2}M_F|H_Q|\frac{5}{2}M_F\rangle = -eqQ/20$.

We approach this problem by treating only the major splitting, which is caused by rotational interaction with just the Cl nuclear spin. We start by evaluating equation Z-5.14.20 with $I = \frac{3}{2}$ and $J = 1$,

$$\Delta E = -\frac{1}{2}eqQ \left[\frac{\frac{3}{4}C(C+1)-I(I+1)J(J+1)}{I(2I-1)(2J+3)(2J-1)}\right]$$

$$= -\frac{eqQ}{30} \left[\frac{3}{4}C(C+1) - \frac{15}{2}\right], \qquad (5.252)$$

to calculate ΔE for each value of the total angular momentum.

For $F = \frac{5}{2}$, equation 5.247 gives $C = 3$. Substituting this value into equation 5.252 gives

$$\Delta E = -\frac{1}{20} eqQ. \qquad (5.253)$$

For $F = \frac{3}{2}$, equation 5.247 gives $C = -2$, and equation 5.252 gives

$$\Delta E = \frac{1}{5} eqQ. \qquad (5.254)$$

Finally, for $F = \frac{1}{2}$, equation 5.247 gives $C = -5$, and equation 5.252 gives

$$\Delta E = -\frac{1}{4} eqQ. \qquad (5.255)$$

To extract a value of eqQ from the data, we first need to average the observed transition energies over the splitting caused by the spin of the nitrogen nucleus. Accordingly, we calculate from the data on page Z-244:

$$\Delta E \left(F = \frac{3}{2} \right) = \frac{1}{3} [11,924.33 + 11,924.89 + 11,925.57]$$

$$= 11,924.93 \; MHz, \tag{5.256}$$

$$\Delta E \left(F = \frac{5}{2} \right) = \frac{1}{3} [11,945.27 + 11,946.06 + 11,946.32]$$

$$= 11,945.88 \; MHz, \tag{5.257}$$

and

$$\Delta E \left(F = \frac{1}{2} \right) = 11,962.56 \; MHz. \tag{5.258}$$

We can use these average transition energies to obtain two independent estimates of eqQ:

$$\Delta E \left(F = \frac{3}{2} \right) - \Delta E \left(F = \frac{1}{2} \right) = \frac{1}{5} eqQ + \frac{1}{4} eqQ$$

$$11,924.93 \; MHz \; - 11,962.56 \; MHz = \frac{9}{20} eqQ$$

$$eqQ = -83.6 MHz, \tag{5.259}$$

and

$$\Delta E \left(F = \frac{5}{2} \right) - \Delta E \left(F = \frac{3}{2} \right) = -\frac{1}{20} eqQ - \frac{1}{5} eqQ$$

$$11,945.88 \; MHz - 11,924.93 \; MHz = -\frac{1}{4} eqQ$$

$$eqQ = -83.8 MHz. \tag{5.260}$$

E.

SET UP, BUT DO NOT SOLVE, THE 2×2 SECULAR DETERMINANT FOR THE $F = \frac{1}{2}$ BLOCK. *Hint*: EVALUATE EQUATION Z-5.14.21 TO SHOW THAT THE NONVANISHING MATRIX ELEMENTS IN THIS BLOCK ARE

$$\left\langle \frac{1}{2} \; 1 \; \frac{1}{2} \middle| H_Q \middle| \frac{1}{2} \; 1 \; \frac{1}{2} \right\rangle = \frac{-eqQ(^{35}Cl)}{4}$$

$$\left\langle \frac{1}{2}\; 1\; \frac{1}{2} \middle| H_Q \middle| \frac{3}{2}\; 1\; \frac{1}{2} \right\rangle = -\frac{\sqrt{5}\,eqQ(^{14}\mathrm{N})}{20}$$

$$= \left\langle \frac{3}{2}\; 1\; \frac{1}{2} \middle| H_Q \middle| \frac{1}{2}\; 1\; \frac{1}{2} \right\rangle$$

$$\left\langle \frac{3}{2}\; 1\; \frac{1}{2} \middle| H_Q \middle| \frac{3}{2}\; 1\; \frac{1}{2} \right\rangle = \frac{eqQ(^{35}\mathrm{Cl}) + eqQ(^{14}\mathrm{N})}{5}. \tag{5.261}$$

For $F = \frac{1}{2}$, the possible values of F_1 that combine with $I_2 = 1$ are $\frac{1}{2}$ and $\frac{3}{2}$. Using standard angular momentum coupling notation,

$$\langle (JI_1)F_1'\; I_2\; F\; M_F | H_Q | (J\; I_1)F_1\; I_2\; F\; M_F \rangle,$$

our task is to evaluate the following four matrix elements:

$$\left\langle \left(1\; \frac{3}{2}\right)\; \frac{1}{2}\; 1\; \frac{1}{2}\; M_F \middle| H_Q \middle| \left(1\; \frac{3}{2}\right)\; \frac{1}{2}\; 1\; \frac{1}{2}\; M_F \right\rangle, \tag{5.262}$$

$$\left\langle \left(1\; \frac{3}{2}\right)\; \frac{1}{2}\; 1\; \frac{1}{2}\; M_F \middle| H_Q \middle| \left(1\; \frac{3}{2}\right)\; \frac{3}{2}\; 1\; \frac{1}{2}\; M_F \right\rangle, \tag{5.263}$$

$$\left\langle \left(1\; \frac{3}{2}\right)\; \frac{3}{2}\; 1\; \frac{1}{2}\; M_F \middle| H_Q \middle| \left(1\; \frac{3}{2}\right)\; \frac{1}{2}\; 1\; \frac{1}{2}\; M_F \right\rangle, \tag{5.264}$$

and

$$\left\langle \left(1\; \frac{3}{2}\right)\; \frac{3}{2}\; 1\; \frac{1}{2}\; M_F \middle| H_Q \middle| \left(1\; \frac{3}{2}\right)\; \frac{3}{2}\; 1\; \frac{1}{2}\; M_F \right\rangle. \tag{5.265}$$

We will evaluate only the matrix element 5.265 explicitly to illustrate the general procedure.

$$\left\langle \left(1\; \frac{3}{2}\right)\; \frac{3}{2}\; 1\; \frac{1}{2}M_F \middle| H_Q \middle| \left(1\; \frac{3}{2}\right)\; \frac{3}{2}\; 1\; \frac{1}{2}M_F \right\rangle$$

$$= \left\langle \left(1\; \frac{3}{2}\right)\; \frac{3}{2}\; 1\; \frac{1}{2}M_F \middle| \mathbf{V}^{(2)}(^{35}\mathrm{Cl}) \cdot \mathbf{Q}^{(2)}(^{35}\mathrm{Cl}) \middle| \left(1\; \frac{3}{2}\right)\; \frac{3}{2}\; 1\; \frac{1}{2}M_F \right\rangle$$

$$+ \left\langle \left(1\; \frac{3}{2}\right)\; \frac{3}{2}\; 1\; \frac{1}{2}M_F \middle| \mathbf{V}^{(2)}(^{14}\mathrm{N}) \cdot \mathbf{Q}^{(2)}(^{14}\mathrm{N}) \middle| \left(1\; \frac{3}{2}\right)\; \frac{3}{2}\; 1\; \frac{1}{2}M_F \right\rangle. \tag{5.266}$$

The first term on the right side becomes, using equation Z-5.64,

$$\left\langle \left(1\; \frac{3}{2}\right)\; \frac{3}{2}\; 1\; \frac{1}{2}M_F \middle| \mathbf{V}^{(2)}(^{35}\mathrm{Cl}) \cdot \mathbf{Q}^{(2)}(^{35}\mathrm{Cl}) \middle| \left(1\; \frac{3}{2}\right)\; \frac{3}{2}\; 1\; \frac{1}{2}M_F \right\rangle$$

$$= (-1)^{\frac{1}{2}-M_F} \begin{pmatrix} \frac{1}{2} & 0 & \frac{1}{2} \\ -M_F & 0 & M_F \end{pmatrix} \left\langle \left(1\,\frac{3}{2}\right)\frac{3}{2}1\,\frac{1}{2} \left\| \mathbf{V}^{(2)}(^{35}\mathrm{Cl}) \cdot \mathbf{Q}^{(2)}(^{35}\mathrm{Cl}) \right\| \left(1\,\frac{3}{2}\right)\frac{3}{2}1\,\frac{1}{2} \right\rangle$$

$$= \frac{1}{\sqrt{2}} \left\langle \left(1\,\frac{3}{2}\right)\frac{3}{2}1\,\frac{1}{2} \left\| \mathbf{V}^{(2)}(^{35}\mathrm{Cl}) \cdot \mathbf{Q}^{(2)}(^{35}\mathrm{Cl}) \right\| \left(1\,\frac{3}{2}\right)\frac{3}{2}1\,\frac{1}{2} \right\rangle. \tag{5.267}$$

With the help of equation Z-5.72, we can further evaluate equation 5.267,

$$\frac{1}{\sqrt{2}} \left\langle \left(1\,\frac{3}{2}\right)\frac{3}{2}1\,\frac{1}{2} \left\| \mathbf{V}^{(2)}(^{35}\mathrm{Cl}) \cdot \mathbf{Q}^{(2)}(^{35}\mathrm{Cl}) \right\| \left(1\,\frac{3}{2}\right)\frac{3}{2}1\,\frac{1}{2} \right\rangle$$

$$= \frac{1}{\sqrt{2}}\,(-1)\,2 \left\{ \begin{matrix} \frac{3}{2} & \frac{1}{2} & 1 \\ \frac{1}{2} & \frac{3}{2} & 0 \\ \frac{1}{2} & \frac{1}{2} & \end{matrix} \right\} \left\langle \left(1\,\frac{3}{2}\right)\frac{3}{2} \left\| \mathbf{V}^{(2)}(^{35}\mathrm{Cl}) \cdot \mathbf{Q}^{(2)}(^{35}\mathrm{Cl}) \right\| \left(1\,\frac{3}{2}\right)\frac{3}{2} \right\rangle$$

$$= \frac{1}{2} \left\langle \left(1\,\frac{3}{2}\right)\frac{3}{2} \left\| \mathbf{V}^{(2)}(^{35}\mathrm{Cl}) \cdot \mathbf{Q}^{(2)}(^{35}\mathrm{Cl}) \right\| \left(1\,\frac{3}{2}\right)\frac{3}{2} \right\rangle. \tag{5.268}$$

The reduced matrix element in equation 5.268 can be calculated using the Wigner-Eckart theorem and equations 5.248 and 5.251 to give

$$\left\langle \left(1\,\frac{3}{2}\right)\frac{3}{2} \left\| \mathbf{V}^{(2)}(^{35}\mathrm{Cl}) \cdot \mathbf{Q}^{(2)}(^{35}\mathrm{Cl}) \right\| \left(1\,\frac{3}{2}\right)\frac{3}{2} \right\rangle$$

$$= (-1)^{\frac{3}{2}-M_{F_1}} \left[\begin{pmatrix} \frac{3}{2} & 0 & \frac{3}{2} \\ -M_{F_1} & 0 & M_{F_1} \end{pmatrix} \right]^{-1}$$

$$\times \left\langle \left(1\,\frac{3}{2}\right)\frac{3}{2}M_{F_1} \left| \mathbf{V}^{(2)}(^{35}\mathrm{Cl}) \cdot \mathbf{Q}^{(2)}(^{35}\mathrm{Cl}) \right| \left(1\,\frac{3}{2}\right)\frac{3}{2}M_{F_1} \right\rangle$$

$$= 2 \times \frac{1}{2}\,eq_{ZZ}(^{35}\mathrm{Cl})Q(^{35}\mathrm{Cl}) \frac{\frac{3}{4}(-2)(-1)-\frac{15}{2}}{3}$$

$$= -2eq_{ZZ}(^{35}\mathrm{Cl})Q(^{35}\mathrm{Cl}). \tag{5.269}$$

Inserting equation 5.269 into 5.268 and using Z-5.14.19, we obtain

$$\left\langle \left(1\,\frac{3}{2}\right)\frac{3}{2}1\,\frac{1}{2}M_F \left| \mathbf{V}^{(2)}(^{35}\mathrm{Cl}) \cdot \mathbf{Q}^{(2)}(^{35}\mathrm{Cl}) \right| \left(1\,\frac{3}{2}\right)\frac{3}{2}1\,\frac{1}{2}M_F \right\rangle = \frac{1}{5}eq(^{35}\mathrm{Cl})Q(^{35}\mathrm{Cl}). \tag{5.270}$$

The second term on the right side of equation 5.266 requires additional attention because $\mathbf{I}_1$ and $\mathbf{F}_1$ do not operate on $\mathbf{V}^{(2)}(^{14}\mathrm{N}) \cdot \mathbf{Q}^{(2)}(^{14}\mathrm{N})$. We proceed by introducing a recoupling transformation (equations

Z-4.3 and Z-4.8),

$$|(J\ I_1)\ F_1\ I_2\ F\ M_F\rangle = \sum_{F_2} |(J\ I_2)\ F_2\ I_1\ F\ M_F\rangle \langle (J\ I_1)\ F_1\ I_2\ F\ M_F| (J\ I_2)\ F_2\ I_1\ F\ M_F\rangle$$

$$= \sum_{F_2} (-1)^{I_1+J+I_2+F} \sqrt{(2F_1+1)(2F_2+1)}$$

$$\times \left\{ \begin{array}{ccc} I_1 & J & F_1 \\ I_2 & F & F_2 \end{array} \right\} |(J\ I_2)\ F_2\ I_1\ F\ M_F\rangle, \tag{5.271}$$

where $\mathbf{F}_2 = \mathbf{J} + \mathbf{I}_2$. Using equation 5.271, we obtain

$$\left|\left(1\ \frac{3}{2}\right)\ \frac{3}{2}\ 1\ \frac{1}{2} M_F\right\rangle = 2 \sum_{F_2} \sqrt{2F_2+1} \left\{ \begin{array}{ccc} \frac{3}{2} & 1 & \frac{3}{2} \\ 1 & \frac{1}{2} & F_2 \end{array} \right\} \left|(11)\ F_2\ \frac{3}{2}\ \frac{1}{2} M_F\right\rangle, \tag{5.272}$$

where F_2 can be $0, 1$, or 2. Using Table Z-4.1 to evaluate the 6-j symbols, we obtain

$$\left|\left(1\ \frac{3}{2}\right)\ \frac{3}{2}\ 1\ \frac{1}{2} M_F\right\rangle = \sqrt{\frac{5}{6}}\left|(11)\ 1\ \frac{3}{2}\ \frac{1}{2} M_F\right\rangle + \frac{1}{\sqrt{6}}\left|(11)\ 2\ \frac{3}{2}\ \frac{1}{2} M_F\right\rangle. \tag{5.273}$$

Using equation 5.273, we obtain for the second term in equation 5.266

$$\left\langle \left(1\ \frac{3}{2}\right)\ \frac{3}{2}\ 1\ \frac{1}{2} M_F \left| \mathbf{V}^{(2)}(^{14}\mathrm{N}) \cdot \mathbf{Q}^{(2)}(^{14}\mathrm{N}) \right| \left(1\ \frac{3}{2}\right)\ \frac{3}{2}\ 1\ \frac{1}{2} M_F \right\rangle$$

$$= \frac{5}{6} \left\langle (11)\ 1\ \frac{3}{2}\ \frac{1}{2} M_F \left| \mathbf{V}^{(2)}(^{14}\mathrm{N}) \cdot \mathbf{Q}^{(2)}(^{14}\mathrm{N}) \right| (11)\ 1\ \frac{3}{2}\ \frac{1}{2} M_F \right\rangle$$

$$+ \frac{1}{6} \left\langle (11)\ 2\ \frac{3}{2}\ \frac{1}{2} M_F \left| \mathbf{V}^{(2)}(^{14}\mathrm{N}) \cdot \mathbf{Q}^{(2)}(^{14}\mathrm{N}) \right| (11)\ 2\ \frac{3}{2}\ \frac{1}{2} M_F \right\rangle$$

$$+ \frac{\sqrt{5}}{3} \left\langle (11)\ 1\ \frac{3}{2}\ \frac{1}{2} M_F \left| \mathbf{V}^{(2)}(^{14}\mathrm{N}) \cdot \mathbf{Q}^{(2)}(^{14}\mathrm{N}) \right| (11)\ 2\ \frac{3}{2}\ \frac{1}{2} M_F \right\rangle, \tag{5.274}$$

where we have used the fact that the two off-diagonal terms are equal. The three matrix elements on the right side of equation 5.274 can be evaluated as before. The results are

$$\left\langle (11)\ 1\ \frac{3}{2}\ \frac{1}{2} M_F \left| \mathbf{V}^{(2)}(^{14}\mathrm{N}) \cdot \mathbf{Q}^{(2)}(^{14}\mathrm{N}) \right| (11)\ 1\ \frac{3}{2}\ \frac{1}{2} M_F \right\rangle = \frac{1}{4} eq(^{14}\mathrm{N})Q(^{14}\mathrm{N}), \tag{5.275}$$

$$\left\langle (11)\ 2\ \frac{3}{2}\ \frac{1}{2} M_F \left| \mathbf{V}^{(2)}(^{14}\mathrm{N}) \cdot \mathbf{Q}^{(2)}(^{14}\mathrm{N}) \right| (11)\ 2\ \frac{3}{2}\ \frac{1}{2} M_F \right\rangle = -\frac{1}{20} eq(^{14}\mathrm{N})Q(^{14}\mathrm{N}), \tag{5.276}$$

and

$$\left\langle (11)\ 1\ \frac{3}{2}\ \frac{1}{2} M_F \left| \mathbf{V}^{(2)}(^{14}\mathrm{N}) \cdot \mathbf{Q}^{(2)}(^{14}\mathrm{N}) \right| (11)\ 2\ \frac{3}{2}\ \frac{1}{2} M_F \right\rangle = 0. \tag{5.277}$$

The last matrix element is zero because the scalar tensor $\mathbf{V}^{(2)}(^{14}\mathrm{N}) \cdot \mathbf{Q}^{(2)}(^{14}\mathrm{N})$ cannot mix two levels with different F_2.

Inserting equations 5.275 through 5.277 into equation 5.274,

$$\left\langle \left(1\,\frac{3}{2}\right)\frac{3}{2}\,1\,\frac{1}{2}M_F \left| \mathbf{V}^{(2)}(^{14}\mathrm{N}) \cdot \mathbf{Q}^{(2)}(^{14}\mathrm{N}) \right| \left(1\,\frac{3}{2}\right)\frac{3}{2}\,1\,\frac{1}{2}M_F \right\rangle = \left(\frac{5}{6}\times\frac{1}{4} - \frac{1}{6}\times\frac{1}{20}\right) eq(^{14}\mathrm{N})Q(^{14}\mathrm{N})$$

$$= \frac{1}{5}eq(^{14}\mathrm{N})Q(^{14}\mathrm{N}). \qquad (5.278)$$

Finally, inserting equations 5.270 and 5.278 into equation 5.266,

$$\left\langle \left(1\,\frac{3}{2}\right)\frac{3}{2}\,1\,\frac{1}{2}M_F \left| H_Q \right| \left(1\,\frac{3}{2}\right)\frac{3}{2}\,1\,\frac{1}{2}M_F \right\rangle = \frac{1}{5}\left[eq(^{35}\mathrm{Cl})Q(^{35}\mathrm{Cl}) + eq(^{14}\mathrm{N})Q(^{14}\mathrm{N})\right]. \qquad (5.279)$$

The procedure described in obtaining equation 5.279 can be repeated to evaluate the other matrix elements in equations 5.263 and 5.265. We note that the off-diagonal matrix elements are equal,

$$\left\langle \left(1\,\frac{3}{2}\right)\frac{1}{2}\,1\,\frac{1}{2}\,M_F \left| H_Q \right| \left(1\,\frac{3}{2}\right)\frac{3}{2}\,1\,\frac{1}{2}\,M_F \right\rangle = \left\langle \left(1\,\frac{3}{2}\right)\frac{3}{2}\,1\,\frac{1}{2}\,M_F \left| H_Q \right| \left(1\,\frac{3}{2}\right)\frac{1}{2}\,1\,\frac{1}{2}\,M_F \right\rangle,$$

$$(5.280)$$

because they are real. The secular determinant for the $F = \frac{1}{2}$ block is therefore given by

$$\begin{vmatrix} \langle \frac{1}{2}\,1\,\frac{1}{2}|H_Q|\frac{1}{2}\,1\,\frac{1}{2}\rangle - E & \langle \frac{1}{2}\,1\,\frac{1}{2}|H_Q|\frac{3}{2}\,1\,\frac{1}{2}\rangle \\[2mm] \langle \frac{3}{2}\,1\,\frac{1}{2}|H_Q|\frac{1}{2}\,1\,\frac{1}{2}\rangle & \langle \frac{3}{2}\,1\,\frac{1}{2}|H_Q|\frac{3}{2}\,1\,\frac{1}{2}\rangle - E \end{vmatrix} = 0, \qquad (5.281)$$

which we have shown to become

$$\begin{vmatrix} -\frac{1}{4}eqQ(^{35}\mathrm{Cl}) - E & -\frac{\sqrt{5}}{20}eqQ(^{14}\mathrm{N}) \\[3mm] -\frac{\sqrt{5}}{20}eqQ(^{14}\mathrm{N}) & \frac{1}{5}\left[eqQ(^{35}\mathrm{Cl}) + eqQ(^{14}\mathrm{N})\right] - E \end{vmatrix} = 0. \qquad (5.282)$$

Chapter 6

ENERGY-LEVEL STRUCTURE AND WAVE FUNCTIONS OF A RIGID ROTOR

PROBLEM SET 4

1.   LET $\hat{\mathbf{n}}$ BE A UNIT VECTOR POINTING ALONG THE ANGULAR VELOCITY VECTOR $\boldsymbol{\omega}$ SO THAT

$$\boldsymbol{\omega} = \omega\hat{\mathbf{n}}. \tag{6.1}$$

THE KINETIC ENERGY OF THE RIGID BODY IN THE CENTER-OF-MASS FRAME IS GIVEN BY

$$T = \frac{1}{2}\boldsymbol{\omega} \cdot \mathbf{J}, \tag{6.2}$$

WHERE

$$\mathbf{J} = \underset{\sim}{\mathbf{I}} \cdot \boldsymbol{\omega}. \tag{6.3}$$

HENCE

$$T = \frac{1}{2}\boldsymbol{\omega} \cdot \underset{\sim}{\mathbf{I}} \cdot \boldsymbol{\omega} = \frac{1}{2}\omega^2\hat{\mathbf{n}} \cdot \underset{\sim}{\mathbf{I}} \cdot \hat{\mathbf{n}} = \frac{1}{2}I\omega^2, \tag{6.4}$$

WHERE I IS THE MOMENT OF INERTIA ABOUT THE AXIS OF ROTATION, A SCALAR QUANTITY GIVEN BY

$$I = \sum_i m_i \left[r_i^2 - (\mathbf{r}_i \cdot \hat{\mathbf{n}})^2 \right]. \tag{6.5}$$

1A.   EVALUATE $\hat{\mathbf{n}} \cdot \underset{\sim}{\mathbf{I}} \cdot \hat{\mathbf{n}}$ AND SHOW THAT EQUATION 6.5 RESULTS.

From equation Z-6.13 we know that the elements of the inertia tensor $\underset{\sim}{\mathbf{I}}$ are given by

$$I_{jk} = \sum_i m_i(r_i^2 \delta_{jk} - r_{ij}r_{ik}). \tag{6.6}$$

We evaluate $\hat{\mathbf{n}} \cdot \underset{\sim}{\mathbf{I}} \cdot \hat{\mathbf{n}}$ explicitly by using equation 6.6:

$$
\begin{aligned}
\hat{\mathbf{n}} \cdot \underset{\sim}{\mathbf{I}} \cdot \hat{\mathbf{n}} &= \sum_{j,k} n_j I_{jk} n_k \\[2mm]
&= \sum_{i,j,k} m_i(r_i^2 \, \delta_{jk} n_j n_k - r_{ij} n_j r_{ik} n_k) \\[2mm]
&= \sum_i m_i r_i^2 \sum_{j,k} \delta_{jk} n_j n_k - \sum_i m_i \sum_j r_{ij} n_j \sum_k r_{ik} n_k \\[2mm]
&= \sum_i m_i r_i^2 - \sum_i m_i (\mathbf{r}_i \cdot \hat{\mathbf{n}})^2 \\[2mm]
&= \sum_i m_i[r_i^2 - (\mathbf{r}_i \cdot \hat{\mathbf{n}})^2].
\end{aligned} \tag{6.7}
$$

1B.   THE DISTANCE OF PARTICLE i FROM THE AXIS OF ROTATION $\hat{\mathbf{n}}$ IS $r_i \sin \theta_i$, WHERE θ_i IS THE ANGLE INCLUDED BETWEEN $\mathbf{r}_i$ AND $\hat{\mathbf{n}}$. THUS THIS DISTANCE IS THE SAME AS THE MAGNITUDE OF THE VECTOR $\mathbf{r}_i \times \hat{\mathbf{n}}$. FOR A RIGID BODY,

$$I = \sum_i m_i (\mathbf{r}_i \times \hat{\mathbf{n}}) \cdot (\mathbf{r}_i \times \hat{\mathbf{n}}). \tag{6.8}$$

SHOW THAT EQUATION 6.8 IS EQUIVALENT TO EQUATION 6.5.

From vector algebra we know that the jth component of the vector $\mathbf{r}_i \times \hat{\mathbf{n}}$ is given by

$$(\mathbf{r}_i \times \hat{\mathbf{n}})_j = \sum_{k,l} \epsilon_{jkl} \, r_{ik} \, n_l. \tag{6.9}$$

Here, ϵ_{jkl} is the permutation symbol, which has the following properties:

$$\epsilon_{jkl} \quad = 0 \quad \text{if any two indices are equal,}$$

$$= 1 \quad \text{if } j, k, l \text{ form an even permutation of 1, 2, 3,}$$

$$= -1 \quad \text{if } j, k, l \text{ form an odd permutation of 1, 2, 3,}$$

and

$$\sum_j \epsilon_{jkl} \, \epsilon_{jpq} = \delta_{kp} \, \delta_{lq} - \delta_{kq} \, \delta_{lp}. \tag{6.10}$$

Inserting equation 6.9 into equation 6.8 and using equation 6.10 yields

$$
\begin{aligned}
I &= \sum_i m_i \, (\mathbf{r}_i \times \hat{\mathbf{n}}) \cdot (\mathbf{r}_i \times \hat{\mathbf{n}}) \\[2ex]
&= \sum_i m_i \sum_j \left[\sum_{k,l} \epsilon_{jkl} \, r_{ik} \, n_l \right] \left[\sum_{p,q} \epsilon_{jpq} \, r_{ip} \, n_q \right] \\[2ex]
&= \sum_i m_i \sum_{klpq} \sum_j \epsilon_{jkl} \, \epsilon_{jpq} \, r_{ik} \, r_{ip} \, n_l n_q \\[2ex]
&= \sum_i m_i \sum_{klpq} \left(\delta_{kp} \, \delta_{lq} - \delta_{kq} \, \delta_{lp} \right) r_{ik} \, r_{ip} \, n_l n_q \\[2ex]
&= \sum_i m_i \sum_{kl} \left[r_{ik} \, r_{ik} n_l n_l - r_{ik} n_k \, r_{il} n_l \right] \\[2ex]
&= \sum_i m_i \left[r_i^2 - (\mathbf{r} \cdot \hat{\mathbf{n}})^2 \right], \tag{6.11}
\end{aligned}
$$

which is the desired result.

2. THE FOLLOWING PROBLEMS RELATE TO THE MICROWAVE SPECTRUM OF METHYLENE FLUORIDE.

2A. CALCULATE THE PRINCIPAL MOMENTS OF INERTIA (EXPRESSED IN AMU $Å^2$) AND ROTATION CONSTANTS A, B, C (EXPRESSED IN MHz AND IN CM^{-1}) FOR THE METHYLENE FLUORIDE MOLECULE, $^{12}CH_2F_2$. USE THE FOLLOWING PARAMETERS:

$r_{CH} = 1.092$ Å, $r_{CF} = 1.358$ Å,

$\angle HCH = 111°52'$, $\angle FCF = 108°17'$,

$m_H = 1.007825$ amu, $m_C = 12.000000$ amu, $m_F = 18.99840$ amu.

A HANDY CONVERSION FACTOR IS $\hbar/4\pi = 5.05377 \times 10^5$ amu $Å^2$ MHz.

To calculate three principal moments of inertia, we should first find the center of mass of the system. In the coordinate system depicted in Figure 6.1, the (x, y, z) coordinates (in Å) for each particle are

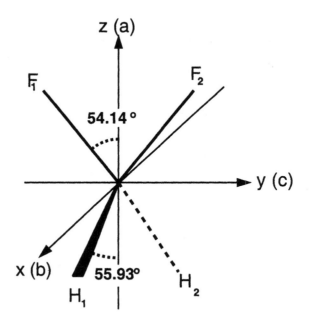

Figure 6.1: Coordinate system used for evaluation of CH_2F_2 parameters.

C : (0, 0, 0)

H_1 : (0.904598, 0, -0.611692)

$H_2 : (-0.904598,\ 0,\ -0.611692)$

$F_1 : (0,\ -1.10062,\ 0.795493).$

$F_2 : (0,\ 1.10062,\ 0.795493).$

The C_{2v} symmetry dictates that the center of mass (c.m.) of the molecule should lie on the z axis.

$$z_{cm} = \frac{\sum_i m_i z_i}{\sum_i m_i}$$

$$= \frac{28.99323}{52.01245}\ \text{Å}$$

$$= 0.557429\ \text{Å}. \tag{6.12}$$

The coordinates of the c.m. are, therefore, $(0,\ 0,\ 0.557429)$. Once we know the c.m. coordinates, it is possible to calculate the components of the moment of inertia tensor using equation Z-6.14. The diagonal elements are

$$I_{xx} = \sum_i m_i \left(\bar{r}_i^2 - \bar{x}_i^2\right) - M\left(R^2 - x_{cm}^2\right)$$

$$= 2 \times 1.007825 \times (1.092^2 - 0.904598^2) + 2 \times 18.99840$$

$$\times (1.358^2 - 0) - 52.01245 \times (0.557429^2 - 0)$$

$$= 54.6648\ \text{amu Å}^2. \tag{6.13}$$

$$I_{yy} = 2 \times 1.007825 \times 1.092^2 + 2 \times 18.99840 \times (1.358^2 - 1.10062^2) - 52.01245 \times 0.557429^2$$

$$= 10.2863\ \text{amu Å}^2, \tag{6.14}$$

and

$$I_{zz} = 2 \times 1.007825 \times (1.092^2 - 0.611692^2) + 2 \times 18.99840 \times (1.358^2 - 0.795493^2)$$

$$= 47.6770\ \text{amu Å}^2. \tag{6.15}$$

The off-diagonal elements are

$$I_{xy} = -\sum_i m_i \bar{x}_i \bar{y}_i + M x_{cm} y_{cm} = 0,$$

$$I_{yz} = -\sum_i m_i \bar{y}_i \bar{z}_i + M y_{cm} z_{cm} = 0,$$

$$I_{zx} = -\sum_i m_i \bar{z}_i \bar{x}_i + M z_{cm} x_{cm} = 0. \tag{6.16}$$

The moment of inertia tensor is given by

$$\underset{\sim}{\mathbf{I}}\,(\text{amu Å}^2) = \begin{bmatrix} 54.6648 & 0 & 0 \\ 0 & 10.2863 & \\ 0 & 0 & 47.6770 \end{bmatrix}. \tag{6.17}$$

The chosen coordinate axes coincide with the principal axes. Following convention, we redesignate the principal axes and the principal moments of inertia as follows:

$$y \to a \qquad I_{aa} = I_{yy} = 10.2863 \text{ amu Å}^2,$$

$$z \to b \qquad I_{bb} = I_{zz} = 47.6770 \text{ amu Å}^2,$$

$$x \to c \qquad I_{cc} = I_{xx} = 54.6648 \text{ amu Å}^2. \tag{6.18}$$

We can evaluate the rotational constants (A, B, C) from the values of the principal moments of inertia:

$$A = \frac{\hbar}{4\pi I_{aa}} = 49131.1 \text{ MHz} = 1.6388 \text{ cm}^{-1}$$

$$B = \frac{\hbar}{4\pi I_{bb}} = 10600.0 \text{ MHz} = 0.3536 \text{ cm}^{-1}$$

$$C = \frac{\hbar}{4\pi I_{cc}} = 9245.0 \text{ MHz} = 0.3084 \text{ cm}^{-1}.$$

2B. LIST THE FREQUENCIES (IN MHZ) OF THE *allowed* ROTATIONAL TRANSITIONS FOR METHYLENE FLUORIDE WITH $J \leq 2$.

The rotational energy for the J_{K_{-1}, K_1} level can be calculated easily based on the formulas given in Table Z-6.5 The results are listed in Table 6.1.

From symmetry considerations, we see that the dipole moment should lie along the b (z; C_2) axis. Therefore, using the selection rule for the asymmetric top transitions,

$$\mu_b \neq 0 \Rightarrow \Delta J = 0, \pm 1$$

$$\Delta K_{-1} = \pm 1, \pm 3$$

$$\Delta K_1 = \pm 1, \pm 3, \tag{6.19}$$

we can evaluate the frequencies for the allowed transitions, which are listed in Table 6.2.

Table 6.1: Rotational energies for the J_{K_{-1},K_1} level.

Level	Energy (MHz)
0_{00}	0
1_{10}	59731.1
1_{11}	58376.1
1_{01}	19845.0
2_{20}	216404.5
2_{21}	216369.4
2_{11}	100776.1
2_{12}	96711.1
2_{02}	59499.9

Table 6.2: Transition frequencies for the allowed transitions.

Transition	Frequency (MHz)
$1_{11} \leftarrow 0_{00}$	58376.1
$1_{10} \leftarrow 1_{01}$	39886.1
$2_{02} \leftarrow 1_{11}$	1123.8
$2_{20} \leftarrow 1_{11}$	158028.4
$2_{21} \leftarrow 1_{10}$	156638.3
$2_{11} \leftarrow 2_{02}$	41276.2
$2_{21} \leftarrow 2_{12}$	119658.3
$2_{20} \leftarrow 2_{11}$	115628.4
$2_{12} \leftarrow 1_{01}$	76866.1

2C. DISCUSS WHETHER THE OBSERVATION OF THE ABOVE TRANSITIONS COULD BE USED TO DETERMINE THE BOND ANGLES AND BOND LENGTHS IN METHYLENE FLUORIDE. ARE ANY ASSUMPTIONS OR ADDITIONAL INFORMATION NEEDED?

In general, a well-resolved spectrum provides the values of three rotational constants, A, B, and C. To uniquely determine the structure of methylene fluoride, however, we need to know four quantities: r_{CH}, r_{CF}, $\angle HCH$, and $\angle FCF$. Because these quantities cannot be determined uniquely from the rotational constants, other experimental data are required to provide an additional constraint. In principle, this information can be obtained from the rotational spectrum of any of the isotopically substituted species.

3. IN TERMS OF THE ROTATIONAL CONSTANTS A, B, AND C, SHOW THAT THE SYMMETRIC TOP ENERGY LEVELS FOR $J = 3$ HAVE THE FORM GIVEN IN TABLE 6.3. *Hint*: AN INSPECTION OF THE SYMMETRIC FORM OF THE SOLUTIONS SHOULD SAVE YOU SOME TIME; YOU DO NOT NEED TO FIND THE ROOTS OF EACH SECULAR DETERMINANT.

Table 6.3: Asymmetric top energy levels for $J = 3$.

Level[a]	General Formula	Prolate Limit $A > B = C$
3_{03}	$2A + 5B + 5C - 2[4(B-C)^2 + (A-B)(A-C)]^{\frac{1}{2}}$	$12B$
3_{13}	$5A + 2B + 5C - 2[4(A-C)^2 + (A-B)(B-C)]^{\frac{1}{2}}$	$11B + A$
3_{12}	$5A + 5B + 2C - 2[4(A-B)^2 + (A-C)(B-C)]^{\frac{1}{2}}$	$11B + A$
3_{22}	$4A + 4B + 4C$	$8B + 4A$
3_{21}	$2A + 5B + 5C + 2[4(B-C)^2 + (A-B)(A-C)]^{\frac{1}{2}}$	$8B + 4A$
3_{31}	$5A + 2B + 5C + 2[4(A-C)^2 + (A-B)(B-C)]^{\frac{1}{2}}$	$3B + 9A$
3_{30}	$5A + 5B + 2C + 2[4(A-B)^2 + (A-C)(B-C)]^{\frac{1}{2}}$	$3B + 9A$

[a] $J_{K(prolate)K(oblate)}$

For $J = 3$, there are seven levels denoted by 3_{03}, 3_{13}, 3_{12}, 3_{22}, 3_{21}, 3_{31}, and 3_{30}. Following the discussion in the text, we choose to work in the prolate symmetric top basis set (representation I in Table Z-6.2). According to equation Z-6.71 and Table Z-6.4, each level belongs to a specific symmetry designation of the $D_2(V)$ group.

Table 6.4: Symmetry designation for each energy level.

Level	Symmetry Designation	Secular Determinant
3_{03}	B_a	E^+
3_{13}	B_b	O^-
3_{12}	B_c	O^+
3_{22}	A	E^-
3_{21}	B_a	E^+
3_{31}	B_b	O^-
3_{30}	B_c	O^+

Table 6.4 shows that there are two levels that belong to each B symmetry designation, with only the 3_{22} level belonging to the A symmetry designation. Therefore, the energy of the 3_{22} level can be obtained by the evaluation of a 1×1 secular determinant, whereas evaluation of energies for the other levels invloves the solution of 2×2 secular determinants.

$E(3_{22})$

The prolate top symmetry basis function belonging to the A designation is given by

$$|J = 3, K = 2, M, S = 1\rangle = \frac{1}{\sqrt{2}}[|3\ 2\ M\rangle - |3\ -2\ M\rangle].$$ (6.20)

The energy of the 3_{22} level, $E(3_{22})$, is therefore

$$E(3_{22}) = \langle 32M | H_R | 32M \rangle$$

$$= \frac{1}{2} [\langle 32M | - \langle 3-2M |] H_R [|32M\rangle - |3-2M\rangle]$$

$$= \frac{1}{2} [\langle 32M | H_R | 32M \rangle - \langle 32M | H_R | 3-2M \rangle$$

$$- \langle 3-2M | H_R | 32M \rangle + \langle 3-2M | H_R | 3-2M \rangle]$$

$$= \frac{1}{2} (B+C)(12-4) + 4A$$

$$= 4A + 4B + 4C, \tag{6.21}$$

where we have used equations Z-6.68 and Z-6.69.

$E(3_{30})$ and $E(3_{12})$

The prolate top symmetry basis functions belonging to the B_c designation are

$$|J=3, K=3, M, S=0\rangle \quad \text{and} \quad |J=3, K=1, M, S=0\rangle.$$

The energies for the 3_{30} and 3_{12} levels can be obtained by solving the following O^+ secular determinant.

$$\begin{vmatrix} \langle 33M0 | H_R | 33M0 \rangle - E & \langle 33M0 | H_R | 31M0 \rangle \\ \\ \langle 31M0 | H_R | 33M0 \rangle & \langle 31M0 | H_R | 31M0 \rangle - E \end{vmatrix} = 0. \tag{6.22}$$

We start by evaluating each element of the secular determinant:

$$\langle 33M0 | H_R | 33M0 \rangle = \frac{1}{2} [\langle 33M | + \langle 3-3M |] H_R [|33M\rangle + |3-3M\rangle]$$

$$= \frac{1}{2} [\langle 33M | H_R | 33M \rangle + \langle 3-3M | H_R | 33M \rangle$$

$$+ \langle 33M | H_R | 3-3M \rangle + \langle 3-3M | H_R | 3-3M \rangle]$$

$$= \frac{1}{2}(B+C)(12-9)+9A$$

$$= 9A+\frac{3}{2}B+\frac{3}{2}C, \tag{6.23}$$

$$\langle 33M0|H_R|31M0\rangle = \frac{1}{2}\left[\langle 33M|+\langle 3-3M|\right]H_R\left[|31M\rangle+|3-1M\rangle\right]$$

$$= \frac{1}{2}\left[\langle 33M|H_R|31M\rangle+\langle 3-3M|H_R|31M\rangle\right.$$

$$\left.+\langle 33M|H_R|3-1M\rangle+\langle 3-3M|H_R|3-1M\rangle\right]$$

$$= \frac{1}{2}\left[\frac{1}{4}(B-C)(12-6)^{\frac{1}{2}}(12-2)^{\frac{1}{2}}+\frac{1}{4}(B-C)(12-6)^{\frac{1}{2}}(12-2)^{\frac{1}{2}}\right]$$

$$= \frac{\sqrt{15}}{2}(B-C), \tag{6.24}$$

$$\langle 31M0|H_R|33M0\rangle = \frac{\sqrt{15}}{2}(B-C), \tag{6.25}$$

$$\langle 31M0|H_R|31M0\rangle = \frac{1}{2}\left[\langle 31M|+\langle 3-1M|\right]H_R\left[|31M\rangle+|3-1M\rangle\right]$$

$$= \frac{1}{2}\left[\langle 31M|H_R|31M\rangle+\langle 31M|H_R|3-1M\rangle\right.$$

$$\left.+\langle 3-1M|H_R|31M\rangle+\langle 3-1M|H_R|3-1M\rangle\right]$$

$$= \frac{1}{2}(B+C)(12-1)+A+\frac{1}{4}(B-C)(12)$$

$$= A+\frac{17}{2}B+\frac{5}{2}C. \tag{6.26}$$

Inserting equations 6.23 to 6.26 into the determinant in equation 6.22, we obtain

$$\begin{vmatrix} 9A + \frac{3}{2}(B+C) - E & \frac{\sqrt{15}}{2}(B-C) \\ \\ \frac{\sqrt{15}}{2}(B-C) & A + \frac{17}{2}B + \frac{5}{2}C - E \end{vmatrix} = 0. \tag{6.27}$$

Evaluating the determinant yields

$$\begin{aligned} 0 &= \left[E - 9A - \frac{3}{2}(B+C)\right]\left[E - A - \frac{17}{2}B - \frac{5}{2}C\right] - \frac{15}{4}(B-C)^2 \\ \\ &= E^2 - \left(9A + \frac{3}{2}B + \frac{3}{2}C + A + \frac{17}{2}B + \frac{5}{2}C\right)E \\ \\ &\quad + \left(9A + \frac{3}{2}B + \frac{3}{2}C\right)\left(A + \frac{17}{2}B + \frac{5}{2}C\right) - \frac{15}{4}(B-C)^2 \\ \\ &= E^2 - 2\left(5A + 5B + 2C\right)E + \left(9A^2 + 78AB + 24AC + 9B^2 + 24BC\right). \tag{6.28} \end{aligned}$$

Finally, solving for E gives

$$\begin{aligned} E &= (5A + 5B + 2C) \pm \left[(5A + 5B + 2C)^2 - (9A^2 + 78AB + 24AC + 9B^2 + 24BC)\right]^{\frac{1}{2}} \\ \\ &= (5A + 5B + 2C) \pm \left(16A^2 + 16B^2 + 4C^2 - 28AB - 4BC - 4AC\right)^{\frac{1}{2}} \\ \\ &= (5A + 5B + 2C) \pm 2\left[4(A - B)^2 + (A - C)(B - C)\right]^{\frac{1}{2}}. \tag{6.29} \end{aligned}$$

Inspecting the behavior of the roots found in equation 6.29 as $B \to C$ and comparing the resulting expressions with the prolate top energies, we can assign each value in equation 6.29 to E(3_{30}) and E(3_{12}), respectively:

$$E(3_{30}) = 5A + 5B + 2C + 2\left[4(A - B)^2 + (A - C)(B - C)\right]^{\frac{1}{2}} \tag{6.30}$$

and

$$E(3_{12}) = 5A + 5B + 2C - 2\left[4(A - B)^2 + (A - C)(B - C)\right]^{\frac{1}{2}}. \tag{6.31}$$

$E(3_{31})$ and $E(3_{13})$

The prolate top symmetry basis functions with the B_b designation are

$$|J = 3, K = 3, M, S = 1\rangle \text{ and } |J = 3, K = 1, M, S = 1\rangle.$$

Following the same procedure as before, we can show that

$$E(3_{31}) = 5A + 2B + 5C + 2\left[4(A-C)^2 + (A-B)(B-C)\right]^{\frac{1}{2}} \tag{6.32}$$

and

$$E(3_{12}) = 5A + 2B + 5C - 2\left[4(A-C)^2 + (A-B)(B-C)\right]^{\frac{1}{2}}. \tag{6.33}$$

$E(3_{21})$ and $E(3_{03})$

The prolate top symmetry basis functions with the B_a designation are

$$|J = 3, K = 2, M, S = 0\rangle \text{ and } |J = 3, K = 0, M, S = 0\rangle.$$

Following the same procedure as before, we can show that

$$E(3_{21}) = 2A + 5B + 5C + 2\left[4(B-C)^2 + (A-B)(A-C)\right]^{\frac{1}{2}} \tag{6.34}$$

and

$$E(3_{03}) = 2A + 5B + 5C - 2\left[4(B-C)^2 + (A-B)(A-C)\right]^{\frac{1}{2}}. \tag{6.35}$$

4.   CONSIDER A SYMMETRIC TRIATOMIC MOLECULE AB_2 THAT BELONGS TO THE POINT GROUP C_{2v}. LET $r = r_{AB}$ DENOTE THE LENGTH OF THE AB BOND AND $\theta = \theta_{BAB}$ DENOTE THE INCLUDED ANGLE. THE MASSES ARE DESIGNATED BY M_A AND M_B.

   4A.   SHOW THAT THE THREE MOMENTS OF INERTIA ARE GIVEN BY

$$I_1 = 2M_B r_{AB}^2 \sin^2\left(\frac{\theta_{BAB}}{2}\right), \tag{6.36}$$

$$I_2 = \frac{2M_A M_B}{M_A + 2M_B} r_{AB}^2 \cos^2\left(\frac{\theta_{BAB}}{2}\right), \tag{6.37}$$

AND

$$I_3 = 2M_B r_{AB}^2\left[\frac{M_A}{M_A + 2M_B} r_{AB}^2 \cos^2\left(\frac{\theta_{BAB}}{2}\right) + \sin^2\left(\frac{\theta_{BAB}}{2}\right)\right], \tag{6.38}$$

WHERE 1 IS THE PRINCIPAL AXIS IN THE PLANE OF THE MOLECULE THAT BISECTS THE θ_{BAB} ANGLE, 2 IS THE PRINCIPAL AXIS IN THE PLANE OF THE MOLECULE THAT IS PERPENDICULAR TO 1, AND 3 IS THE PRINCIPAL AXIS PERPENDICULAR TO THE PLANE OF THE MOLECULE (I.E., PERPENDICULAR TO AXES 1 AND 2).

We start by evaluating the position of the center of mass of the system. In the coordinate system given in Figure 6.2, the (x, y) coordinates for each particle are given by

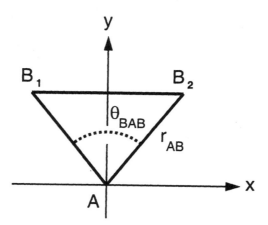

Figure 6.2: Coordinate system for the AB_2 molecule.

A :   (0, 0)

B_1 : $\left(r_{AB} \sin \left(\frac{\theta_{BAB}}{2} \right),\ r_{AB} \cos \left(\frac{\theta_{BAB}}{2} \right) \right)$

B_2 : $\left(-r_{AB} \sin \left(\frac{\theta_{BAB}}{2} \right),\ r_{AB} \cos \left(\frac{\theta_{BAB}}{2} \right) \right).$ (6.39)

Because all three particles are in the same plane, the z coordinate is always zero. The (x, y) coordinates of the center of mass (c.m.) are

$$x_{\text{cm}} = \frac{1}{M} \left[M_B\, r_{AB}\, \sin \left(\frac{\theta_{BAB}}{2} \right) - M_B\, r_{AB}\, \sin \left(\frac{\theta_{BAB}}{2} \right) \right]$$

$$= 0 \tag{6.40}$$

$$y_{\text{cm}} = \frac{2}{M} \left[M_B\, r_{AB}\, \cos \left(\frac{\theta_{BAB}}{2} \right) \right]$$

$$= R, \tag{6.41}$$

where $M = M_A + 2M_B$. Having established these coordinates, we can deduce the components of the moment of inertia tensor, using equation Z-6.14:

$$I_{xx} = M_A(0 - 0) + 2M_B \left[r_{AB}^2 - r_{AB}^2\, \sin^2 \left(\frac{\theta_{BAB}}{2} \right) \right] - MR^2$$

$$= 2M_B\, r_{AB}^2 \cos^2\left(\frac{\theta_{BAB}}{2}\right) - \frac{4M_B^2}{M} r_{AB}^2 \cos^2\left(\frac{\theta_{BAB}}{2}\right)$$

$$= \frac{2M_A M_B}{M}\, r_{AB}^2 \cos^2\left(\frac{\theta_{BAB}}{2}\right), \tag{6.42}$$

$$I_{yy} = M_A(0-0) + 2M_B\left[r_{AB}^2 - r_{AB}^2 \cos^2\left(\frac{\theta_{BAB}}{2}\right)\right]$$

$$= 2M_B\, r_{AB}^2 \sin^2\left(\frac{\theta_{BAB}}{2}\right), \tag{6.43}$$

$$I_{zz} = 2M_B\, r_{AB}^2 - MR^2$$

$$= 2M_B\, r_{AB}^2 - \frac{4M_B^2}{M} r_{AB}^2 \cos^2\left(\frac{\theta_{BAB}}{2}\right)$$

$$= 2M_B\, r_{AB}^2 \left[\frac{M_A}{M} \cos^2\left(\frac{\theta_{BAB}}{2}\right) + \sin^2\left(\frac{\theta_{BAB}}{2}\right)\right], \tag{6.44}$$

$$I_{xy} = I_{yx}$$

$$= -M_B\, r_{AB}^2\, \cos\left(\frac{\theta_{BAB}}{2}\right)\, \sin\left(\frac{\theta_{BAB}}{2}\right)$$

$$\quad + M_B\, r_{AB}^2\, \cos\left(\frac{\theta_{BAB}}{2}\right)\, \sin\left(\frac{\theta_{BAB}}{2}\right) + M \times 0$$

$$= 0, \tag{6.45}$$

$$I_{yz} = I_{zy} = 0, \tag{6.46}$$

$$I_{zx} = I_{xz} = 0. \tag{6.47}$$

From equations 6.42 to 6.47, we conclude that the three principal moments of inertia are given by

$$I_1 = I_{yy} = 2M_B\, r_{AB}^2\, \sin^2\left(\frac{\theta_{BAB}}{2}\right), \tag{6.48}$$

$$I_2 = I_{xx} = \frac{2M_B M_A}{M} r_{AB}^2\, \cos^2\left(\frac{\theta_{BAB}}{2}\right), \tag{6.49}$$

and

$$I_3 = I_{zz} = 2M_B r_{AB}^2 \left[\frac{M_A}{M} \cos^2 \left(\frac{\theta_{BAB}}{2} \right) + \sin^2 \left(\frac{\theta_{BAB}}{2} \right) \right].$$ (6.50)

For I_1 to equal I_2, it is necessary that

$$\theta_{BAB} = 2 \arctan \left(\frac{M_A}{M} \right)^{\frac{1}{2}}$$

$$= \theta_{BAB}^0.$$ (6.51)

For $\theta_{BAB} > \theta_{BAB}^0$, $I_{aa} = I_2$ and $I_{bb} = I_1$, whereas for $\theta_{BAB} < \theta_{BAB}^0$, $I_{aa} = I_1$ and $I_{bb} = I_2$.

4B. SHOW THAT THE ASYMMETRY PARAMETER κ DEPENDS ONLY ON THE MASS RATIO

$$r = \frac{M_A}{M_A + 2M_B}.$$ (6.52)

Equation Z-6.80 defines the asymmetry parameter κ as

$$\kappa = \frac{2B - A - C}{A - C}$$

$$= \frac{\frac{2}{I_{bb}} - \frac{1}{I_{aa}} - \frac{1}{I_{cc}}}{\frac{1}{I_{aa}} - \frac{1}{I_{cc}}}$$

$$= \frac{2I_{aa}I_{cc} - I_{bb}I_{cc} - I_{aa}I_{bb}}{I_{bb}I_{cc} - I_{aa}I_{bb}}.$$ (6.53)

For $\theta_{BAB} < \theta_{BAB}^0$, the principal moments of inertia are given by

$$I_{aa} = 2M_B r_{AB}^2 \sin^2 \left(\frac{\theta_{BAB}}{2} \right),$$ (6.54)

$$I_{bb} = 2 \frac{M_A M_B}{M} r_{AB}^2 \cos^2 \left(\frac{\theta_{BAB}}{2} \right),$$ (6.55)

and

$$I_{cc} = 2M_B r_{AB}^2 \left[\frac{M_A}{M} \cos^2 \left(\frac{\theta_{BAB}}{2} \right) + \sin^2 \left(\frac{\theta_{BAB}}{2} \right) \right].$$ (6.56)

It follows that

$$I_{aa}I_{bb} = \frac{4M_A M_B^2}{M} r_{AB}^4 \sin^2 \left(\frac{\theta_{BAB}}{2} \right) \cos^2 \left(\frac{\theta_{BAB}}{2} \right),$$ (6.57)

$$I_{bb}I_{cc} \;=\; \frac{4M_A M_B^2}{M} r_{AB}^4 \left[\frac{M_A}{M} \cos^4\left(\frac{\theta_{BAB}}{2}\right) + \sin^2\left(\frac{\theta_{BAB}}{2}\right)\cos^2\left(\frac{\theta_{BAB}}{2}\right) \right], \tag{6.58}$$

and

$$I_{cc}I_{aa} \;=\; 4M_B^2 r_{AB}^4 \left[\frac{M_A}{M} \sin^2\left(\frac{\theta_{BAB}}{2}\right) \cos^2\left(\frac{\theta_{BAB}}{2}\right) + \sin^4\left(\frac{\theta_{BAB}}{2}\right) \right]. \tag{6.59}$$

By inserting equations 6.57 through 6.59 into equation 6.53, we obtain for κ

$$\kappa \;=\; \frac{8M_B^2 \sin^4\left(\frac{\theta_{BAB}}{2}\right)}{\frac{4M_A^2 M_B^2}{M^2}\cos^4\left(\frac{\theta_{BAB}}{2}\right)} - 1$$

$$\;=\; \frac{2\tan^4\left(\frac{\theta_{BAB}}{2}\right)}{r^2} - 1, \tag{6.60}$$

where

$$r \;=\; \frac{M_A}{M}$$

$$\;=\; \tan^2\left(\frac{\theta_{BAB}^0}{2}\right). \tag{6.61}$$

For $\theta_{BAB} > \theta_{BAB}^0$, we can follow the same procedure, but now we reverse the designations of I_{aa} and I_{bb}. We obtain for κ

$$\kappa = 2r^2 \cot^4\left(\frac{\theta_{BAB}}{2}\right) - 1. \tag{6.62}$$

We conclude that κ is a function of r and θ_{BAB} only.

4C.   MAKE A PLOT OF κ VERSUS θ_{BAB} FOR $r = \frac{1}{3}$ (CORRESPONDING TO $M_A = M_B$), $r = \frac{1}{2}$ (CORRESPONDING TO $M_A = 2M_B$), AND $r = \frac{1}{5}$ (CORRESPONDING TO $M_A = 0.5M_B$).

From equations 6.60 and 6.62 we see that $\kappa = 1$ for $\theta_{BAB} = \theta_{BAB}^0$. Examination of the graph of κ versus θ_{BAB} in Figure 6.3 shows κ has a value near -1 (which is characteristic of a prolate symmetric top) for all angles far from θ_{BAB}^0.

4D.   WE DEFINE THE INERTIA DEFECT Δ AS

$$\Delta = I_{cc} - I_{bb} - I_{aa}. \tag{6.63}$$

NOTE THAT $\Delta = 0$ FOR A SYMMETRIC TRIATOMIC MOLECULE. PROVE THAT $\Delta = 0$ FOR ANY PLANAR RIGID ROTOR.

For a planar molecule, it is clear from symmetry considerations that one of the principal axes is perpendicular to the molecular plane. We identify this principal axis with our z axis. We also choose the c.m. of

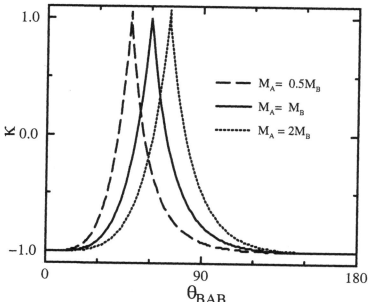

Figure 6.3: Asymmetry parameter for $r = \frac{1}{3}$ ($M_A = M_B$), $r = \frac{1}{2}$ ($M_A = 2M_B$), and $r = \frac{1}{5}$ ($M_A = 0.5M_B$).

the molecule (which lies in the molecular plane) as the origin of the coordinate system. From the definitions of the moments of inertia, we know that

$$I_{xx} = \sum_i m_i \left(r_i^2 - x_i^2 \right), \tag{6.64}$$

$$I_{yy} = \sum_i m_i \left(r_i^2 - y_i^2 \right), \tag{6.65}$$

$$I_{zz} = \sum_i m_i r_i^2. \tag{6.66}$$

Note that the i index runs for all the atoms in the molecule and that $z_i = 0$. Note as well that the x and y axes are *not* defined to be the principal axes yet. Because all the atoms lie in the plane,

$$r_i^2 = x_i^2 + y_i^2. \tag{6.67}$$

From equations 6.64 to 6.67,

$$I_{zz} - I_{yy} - I_{xx} = \sum_i m_i \left[r_i^2 - \left(r_i^2 - y_i^2 \right) - \left(r_i^2 - x_i^2 \right) \right]$$

$$= 0. \tag{6.68}$$

Equation 6.68 should hold for any choice of x and y axes, so long as they are in the plane containing the molecule. Equation 6.68 therefore holds for the principal moments of inertia, proving that

$$\Delta = I_{cc} - I_{bb} - I_{aa} = 0. \tag{6.69}$$

4E.   SUGGEST HOW A REAL "PLANAR" MOLECULE, SUCH AS BENZENE OR OZONE, CAN HAVE A NONVANISHING INERTIA DEFECT.

Real "planar" molecules are nonrigid rotors; therefore, they can deform. In fact, Δ can be represented as the sum of vibrational, centrifugal, and electronic contributions.

$$\Delta = \Delta(\text{vib}) + \Delta(\text{centr}) + \Delta(\text{electr}). \tag{6.70}$$

In most cases, $\Delta(\text{vib})$ dominates in magnitude over the other terms.

APPLICATION 15
INTRODUCTION TO DIATOMIC MOLECULES

A.

USE THE VAN VLECK TRANSFORMATION TO CALCULATE THE FIRST CENTRIFUGAL DISTORTION CORRECTION TO THE ROTATIONAL ENERGY LEVELS OF A $^1\Sigma$ STATE. SHOW THAT

$$E(^1\Sigma; vJ) = B_v J(J+1) - D_v \left[J(J+1)\right]^2 , \tag{6.71}$$

WHERE

$$D_v = - \sum_{v' \neq v} \frac{\langle v|B(r)|v'\rangle \, \langle v'|B(r)|v\rangle}{E_v - E_{v'}} . \tag{6.72}$$

The rotational Hamiltonian is given by

$$\mathcal{H}_{\text{rot}} = B(r)\mathbf{J}^2 . \tag{6.73}$$

The leading term in the rotational energy is

$$
\begin{aligned}
\langle v; JM \,|\mathcal{H}_{\text{rot}}| \, v; JM \rangle &= \langle v|B(r)|v\rangle \, \langle JM|\mathbf{J}^2|JM\rangle \\[2mm]
&= B_v J(J+1) .
\end{aligned}
\tag{6.74}
$$

The second-order correction is given by equation Z-6.104,

$$
\begin{aligned}
\mathcal{H}^{(2)}_{vJ;v'J'} &= \sum_{v'J'M'} \frac{\langle v; JM \,|\mathcal{H}_{\text{rot}}| \, v'; J'M'\rangle \, \langle v'; J'M' \,|\mathcal{H}_{\text{rot}}| \, v; JM\rangle}{E_{vJ} - E_{v'J'}} \\[3mm]
&= \sum_{v \neq v'} \langle v|B(r)|v'\rangle \, \langle v'|B(r)|v\rangle \sum_{J'M'} \frac{\langle JM \,|\mathbf{J}^2| \, J'M'\rangle \, \langle J'M' \,|\mathbf{J}^2| \, JM\rangle}{E_{vJ} - E_{v'J'}} \\[3mm]
&= \sum_{v \neq v'} \sum_{J'M'} \frac{\langle v|B(r)|v'\rangle \, \langle v'|B(r)|v\rangle \, J^2(J+1)^2 \, \delta_{JJ'} \, \delta_{MM'}}{E_{vJ} - E_{v'J'}} \\[3mm]
&= \sum_{v \neq v'} \frac{\langle v|B(r)|v'\rangle \, \langle v'|B(r)|v\rangle \, J^2(J+1)^2}{E_v - E_{v'}} .
\end{aligned}
\tag{6.75}
$$

Combining equations 6.74 and 6.75 gives

$$E(vJ) = B_v J(J+1) - D_v \left[J(J+1)\right]^2 , \tag{6.76}$$

where D_v is defined in equation 6.72.

B.

SHOW THAT

$$E(^2\Sigma; \ vJp^+) \ = \ B_v J\left(J - \frac{1}{2}\right)\left(J + \frac{1}{2}\right) + \frac{1}{2}\gamma_v\left(J - \frac{1}{2}\right)$$

$$= \ B_v N\left(N + 1\right) + \frac{1}{2}\gamma_v N, \tag{6.77}$$

WHERE $J = N + \frac{1}{2}$, AND

$$E(^2\Sigma; \ vJp^-) \ = \ B\left(J + \frac{1}{2}\right)\left(J + \frac{3}{2}\right) - \frac{1}{2}\gamma_v\left(J + \frac{3}{2}\right)$$

$$= \ B_v N\left(N + 1\right) - \frac{1}{2}\gamma_v\left(N + 1\right), \tag{6.78}$$

WHERE $J = N - \frac{1}{2}$. *Hint*: EVALUATE $\left\langle \psi(p^\pm) \left| \mathcal{H}_{\rm rot}^{(v)} \right| \psi(p^\pm) \right\rangle$. USE THE PHASE CONVENTIONS (SEE SECTION Z-3.4) GIVEN IN EQUATIONS Z-6.15.17 TO Z-6.15.22.

The wave functions for Hund's case (a) coupling with $p^\pm$ parity are

$$|\psi(p^\pm)\rangle = \frac{1}{\sqrt{2}}\left[|n\Lambda\rangle \ |S\Sigma\rangle \ |J\Omega M\rangle \pm |n-\Lambda\rangle \ |S-\Sigma\rangle \ |J-\Omega M\rangle\right]. \tag{6.79}$$

We choose $\Sigma = \Omega = \frac{1}{2}$ and calculate a number of useful matrix elements:

$$\left\langle S\frac{1}{2}\left|\left\langle J\frac{1}{2}\right| \mathbf{J}^2 - 2J_z S_z + \mathbf{S}^2 \left|S\frac{1}{2}\right\rangle\right| J\frac{1}{2}\right\rangle \ = \ J(J + 1) - 2\frac{1}{2}\frac{1}{2} + \frac{1}{2}\frac{3}{2}$$

$$= \ J(J + 1) + \frac{1}{4}, \tag{6.80}$$

$$\left\langle S-\frac{1}{2}\left|\left\langle J-\frac{1}{2}\right| \mathbf{J}^2 - 2J_z S_z + \mathbf{S}^2 \left|S-\frac{1}{2}\right\rangle\right| J-\frac{1}{2}\right\rangle \ = \ J(J + 1) + \frac{1}{4}, \tag{6.81}$$

$$\left\langle S\frac{1}{2}\left|\left\langle J\frac{1}{2}\right| \mathbf{J}^2 - 2J_z S_z + \mathbf{S}^2 \left|S-\frac{1}{2}\right\rangle\right| J-\frac{1}{2}\right\rangle = 0, \tag{6.82}$$

$$\left\langle S\frac{1}{2}\left|\left\langle J\frac{1}{2}\right| J_z S_z - \mathbf{S}^2 \left|S\frac{1}{2}\right\rangle\right| J\frac{1}{2}\right\rangle \ = \ -\frac{1}{2}, \tag{6.83}$$

$$\left\langle S-\frac{1}{2}\left|\left\langle J-\frac{1}{2}\right| J_z S_z - \mathbf{S}^2 \left|S-\frac{1}{2}\right\rangle\right| J-\frac{1}{2}\right\rangle = -\frac{1}{2}, \tag{6.84}$$

$$\left\langle S \ -\frac{1}{2}\right|\left\langle J \ -\frac{1}{2}\right| J_z S_z - \mathbf{S}^2 \left|S \ \frac{1}{2}\right\rangle\left| J \ \frac{1}{2}\right\rangle = 0. \tag{6.85}$$

$$\left\langle J \ -\frac{1}{2}\right| J_+ \left| J \ \frac{1}{2}\right\rangle \ = \ \left\langle J \ \frac{1}{2}\right| J_- \left| J \ -\frac{1}{2}\right\rangle$$

$$= \ J + \frac{1}{2}, \tag{6.86}$$

$$\left\langle S \ -\frac{1}{2}\right| S_- \left| S \ \frac{1}{2}\right\rangle \ = \ \left\langle S \ \frac{1}{2}\right| S_+ \left| S \ -\frac{1}{2}\right\rangle$$

$$= \ 1. \tag{6.87}$$

The diagonal matrix elements of $J_\pm$ and $S_\pm$ are all zero. In equation 6.86, the anomalous commutation relations (equation Z-3.38), appropriate for calculating matrix elements of $J_\pm$ in the molecular frame, were used.

We can use these matrix elements to calculate the expectation value of $\mathcal{H}_{\text{rot}}$, given in equation Z-6.15.14. The matrix elements that are proportional to B_v are

$$\left\langle \psi^+ \left| \mathbf{J}^2 - 2J_z S_z + \mathbf{S}^2 - J_+ S_- - J_- S_+ \right| \psi^+ \right\rangle \ = \ J(J+1) + \frac{1}{4} - \left(J + \frac{1}{2}\right)$$

$$= \ \left(J - \frac{1}{2}\right)\left(J + \frac{1}{2}\right) \tag{6.88}$$

and

$$\left\langle \psi^- \left| \mathbf{J}^2 - 2J_z S_z + \mathbf{S}^2 - J_+ S_- - J_- S_+ \right| \psi^- \right\rangle \ = \ J(J+1) + \frac{1}{4} + \left(J + \frac{1}{2}\right)$$

$$= \ \left(J + \frac{1}{2}\right)\left(J + \frac{3}{2}\right). \tag{6.89}$$

The matrix elements proportional to γ_v are

$$\left\langle \psi^+ \left| J_z S_z - \mathbf{S}^2 + \frac{1}{2}J_+ S_- + \frac{1}{2}J_- S_+ \right| \psi^+ \right\rangle \ = \ -\frac{1}{4} + \frac{J}{2} \tag{6.90}$$

and

$$\left\langle \psi^- \left| J_z S_z - \mathbf{S}^2 + \frac{1}{2}J_+ S_- + \frac{1}{2}J_- S_+ \right| \psi^- \right\rangle \ = \ -\frac{3}{4} - \frac{J}{2}. \tag{6.91}$$

Multiplying equation 6.88 by B_v and 6.90 by γ_v and adding, we obtain

$$E(^2\Sigma; \ vJp^+) = B_v\left(J - \frac{1}{2}\right)\left(J + \frac{1}{2}\right) + \frac{1}{2}\gamma_v\left(J - \frac{1}{2}\right), \tag{6.92}$$

whereas multiplying equations 6.89 by B_v and equation 6.91 by γ_v and adding, we obtain

$$E(^2\Sigma;\ vJp^-) = B_v \left(J + \frac{1}{2}\right)\left(J + \frac{3}{2}\right) - \frac{1}{2}\gamma_v\left(J + \frac{3}{2}\right). \tag{6.93}$$

C.

THE ROTATIONAL ENERGY LEVELS OF A $^2\Pi$ STATE ARE THE SAME FOR EACH PARITY BLOCK; THAT IS, $E(^2\Pi;\ vJp^+) = E(^2\Pi;\ vJp^-)$. HENCE, THEY OCCUR IN DOUBLY DEGENERATE PAIRS. SHOW THAT FOR $J > \frac{1}{2}$,

$$E(^2\Pi;\ vJ) = B_v\left[\left(J - \frac{1}{2}\right)\left(J + \frac{3}{2}\right) \pm \frac{1}{2}X\right], \tag{6.94}$$

WHERE

$$X = \left[4\left(J - \frac{1}{2}\right)\left(J + \frac{3}{2}\right) + (Y - 2)^2\right]^{\frac{1}{2}} \tag{6.95}$$

AND

$$Y = \frac{A_v}{B_v}. \tag{6.96}$$

Hint: SHOW THAT THE 2 × 2 SECULAR DETERMINANT IS

$$\begin{vmatrix} B_v\left[J(J+1) - \frac{7}{4}\right] + \frac{1}{2}A_v - E & -B_v\left[J(J+1) - \frac{3}{4}\right]^{\frac{1}{2}} \\ \\ -B_v\left[J(J+1) - \frac{3}{4}\right]^{\frac{1}{2}} & B_v\left[J(J+1) + \frac{1}{4}\right] - \frac{1}{2}A_v - E \end{vmatrix} = 0 \tag{6.97}$$

AND CARRY OUT ITS EVALUATION, USING THE IDENTITY $J(J+1) - \frac{3}{4} = (J - \frac{1}{2})(J + \frac{3}{2})$.

Using the notation of Part B, we define the following wave functions:

$$|\psi_1\rangle = \frac{1}{\sqrt{2}}\left[\left|n\Lambda\right\rangle\left|S\ \frac{1}{2}\right\rangle\left|J\ \frac{3}{2}\right\rangle + \left|n\ -\Lambda\right\rangle\left|S\ -\frac{1}{2}\right\rangle\left|J\ -\frac{3}{2}\right\rangle\right] \tag{6.98}$$

and

$$|\psi_2\rangle = \frac{1}{\sqrt{2}}\left[\left|n\Lambda\right\rangle\left|S\ -\frac{1}{2}\right\rangle\left|J\ \frac{1}{2}\right\rangle + \left|n\ -\Lambda\right\rangle\left|S\ \frac{1}{2}\right\rangle\left|J\ -\frac{1}{2}\right\rangle\right], \tag{6.99}$$

where $\Lambda = 1$ and $S = \frac{1}{2}$.

We want to evaluate the following matrix elements: $\mathcal{H}_{11}$, $\mathcal{H}_{22}$, and $\mathcal{H}_{12} = \mathcal{H}_{21}$. We use equations Z-6.15.17 through Z-6.15.22 to evaluate the matrix elements of $\mathcal{H}_{\text{rot}}$ and $\mathcal{H}_{\text{SO}}$. For $\mathcal{H}_{11}$ and $\mathcal{H}_{22}$, the contributions from $\Lambda = 1$ and $\Lambda = -1$ are each multiplied by $\frac{1}{2}$, and the matrix elements of J_+S_- and J_-S_+ are zero. The results, showing the term-by-term contributions from equations Z-6.15.27 and Z-6.15.29, are

$$\langle \psi_1 | \mathcal{H}_{\text{rot}} | \psi_1 \rangle = B_v \left[J(J+1) - 2\frac{3}{2}\frac{1}{2} + \frac{1}{2}\frac{3}{2} - 2\left(\frac{3}{2} - \frac{1}{2}\right)(1) + 1 \right]$$

$$= B_v \left[J(J+1) - \frac{7}{4} \right], \tag{6.100}$$

$$\langle \psi_1 | \mathcal{H}_{\text{SO}} | \psi_1 \rangle = \frac{1}{2} A_v, \tag{6.101}$$

$$\langle \psi_2 | \mathcal{H}_{\text{rot}} | \psi_2 \rangle = B_v \left\{ J(J+1) - 2\frac{1}{2}\left(-\frac{1}{2}\right) + \frac{1}{2}\frac{3}{2} - 2\left[\frac{1}{2} - \left(-\frac{1}{2}\right)\right] + 1 \right\}$$

$$= B_v \left[J(J+1) + \frac{1}{4} \right], \tag{6.102}$$

and

$$\langle \psi_2 | \mathcal{H}_{\text{SO}} | \psi_2 \rangle = -\frac{1}{2} A_v. \tag{6.103}$$

For the off-diagonal matrix element, only J_+S_- and J_-S_+ have nonvanishing contributions:

$$\langle \psi_1 | \mathcal{H}_{\text{rot}} | \psi_2 \rangle = -B_v \left[J(J+1) - \frac{3}{2}\frac{1}{2} \right]^{\frac{1}{2}} = -B_v \left[\left(J - \frac{1}{2}\right)\left(J + \frac{3}{2}\right) \right]^{\frac{1}{2}}, \tag{6.104}$$

and

$$\langle \psi_1 | \mathcal{H}_{\text{SO}} | \psi_2 \rangle = 0. \tag{6.105}$$

Finally, we obtain the matrix elements in the secular determinant (equation 6.97) by noting that

$$\mathcal{H}_{ij} = \langle \psi_i | \mathcal{H}_{\text{rot}} + \mathcal{H}_{\text{SO}} | \psi_j \rangle. \tag{6.106}$$

Inserting all of the $\mathcal{H}_{ij}$ into the secular determinant, we can solve the secular equation

$$(\mathcal{H}_{11} - E)(\mathcal{H}_{22} - E) - \mathcal{H}_{12}^2 = 0, \tag{6.107}$$

to obtain the eigenvalues

$$E = \frac{1}{2}(\mathcal{H}_{11} + \mathcal{H}_{22}) \pm \frac{1}{2}\left[(\mathcal{H}_{11} - \mathcal{H}_{22})^2 + 4\mathcal{H}_{12}^2\right]^{\frac{1}{2}}, \tag{6.108}$$

where

$$\mathcal{H}_{11} - \mathcal{H}_{22} = A_v - 2B_v, \tag{6.109}$$

and

$$\mathcal{H}_{11} + \mathcal{H}_{22} = 2B_v \left[J(J+1) - \frac{3}{4} \right]. \tag{6.110}$$

The two solutions are

$$E \;=\; B_v \left(J - \frac{1}{2}\right)\left(J + \frac{3}{2}\right) \pm \frac{1}{2}\left[4B_v^2\left(J - \frac{1}{2}\right)\left(J + \frac{3}{2}\right) + (A_v - 2\,B_v)^2\right]^{\frac{1}{2}}$$

$$\;=\; B_v\left\{\left(J - \frac{1}{2}\right)\left(J + \frac{3}{2}\right) \pm \frac{1}{2}\left[4\left(J - \frac{1}{2}\right)\left(J + \frac{3}{2}\right) + \left(\frac{A_v}{B_v} - 2\right)^2\right]^{\frac{1}{2}}\right\}. \tag{6.111}$$

D.

<small>THE ENERGY LEVELS ASSOCIATED WITH THE PLUS SIGN IN EQUATION 6.111 ARE CALLED F_2; THOSE WITH THE MINUS SIGN ARE CALLED F_1. SHOW THAT</small>

$$|\psi(F_2)\rangle = a_J \left|{}^2\Pi_{\frac{3}{2}}vJ\right\rangle - b_J \left|{}^2\Pi_{\frac{1}{2}}vJ\right\rangle \tag{6.112}$$

$$|\psi(F_1)\rangle = b_J \left|{}^2\Pi_{\frac{3}{2}}vJ\right\rangle + a_J \left|{}^2\Pi_{\frac{1}{2}}vJ\right\rangle, \tag{6.113}$$

<small>WHERE</small>

$$a_J = \left[\frac{X + (Y - 2)}{2X}\right]^{\frac{1}{2}} \tag{6.114}$$

<small>AND</small>

$$b_J = \left[\frac{X - (Y - 2)}{2X}\right]^{\frac{1}{2}}. \tag{6.115}$$

From the secular equation, we can solve for the eigenfunctions as follows:

$$\begin{pmatrix} \mathcal{H}_{11} - E & \mathcal{H}_{12} \\ \mathcal{H}_{21} & \mathcal{H}_{22} - E \end{pmatrix} \begin{pmatrix} a_J \\ -b_J \end{pmatrix} = 0. \tag{6.116}$$

For the expansion coefficients of $|\psi(F_2)\rangle$ we have

$$a_J\left(\mathcal{H}_{11} - E_{F_2}\right) - b_J\mathcal{H}_{12} = 0, \tag{6.117}$$

from which it follows that

$$\frac{a_J}{b_J} = \frac{\mathcal{H}_{12}}{\mathcal{H}_{11} - E_{F_2}} \tag{6.118}$$

and

$$\frac{a_J^2}{b_J^2} = \frac{\mathcal{H}_{12}^2}{(\mathcal{H}_{11} - E_{F_2})^2}. \tag{6.119}$$

Defining

$$G = \left(J - \frac{1}{2}\right)\left(J + \frac{3}{2}\right) \tag{6.120}$$

and recalling the definitions of X and Y in equations 6.95 and 6.96, we find that

$$\mathcal{H}_{12} = -B_v \sqrt{G} \qquad (6.121)$$

and

$$\mathcal{H}_{11} - E_{F_2} = B_v \left[G - 1 + \frac{Y}{2} \right] - B_v \left[G + \frac{X}{2} \right]$$

$$= -\frac{1}{2} B_v \left[X - (Y - 2) \right]. \qquad (6.122)$$

Inserting equations 6.121 and 6.122 into 6.119, we obtain

$$\frac{a_J^2}{b_J^2} = \frac{4G}{[X - (Y - 2)]^2}. \qquad (6.123)$$

Solving equation 6.95 for G and substituting the result into equation 6.123 gives

$$\frac{a_J^2}{b_J^2} = \frac{4[X^2 - (Y - 2)^2]/4}{[X - (Y - 2)]^2}$$

$$= \frac{[X + (Y - 2)][X - (Y - 2)]}{[X - (Y - 2)]^2}$$

$$= \frac{X + (Y - 2)}{X - (Y - 2)} \qquad (6.124)$$

Normalization of the eigenvectors requires $a_J^2 + b_J^2 = 1$. Substituting $b_J^2 = 1 - a_J^2$ into equation 6.124 and solving for a_J and b_J yields

$$a_J = \left[\frac{X + (Y - 2)}{2X} \right]^{\frac{1}{2}} \qquad (6.125)$$

and

$$b_J = \left[\frac{X - (Y - 2)}{2X} \right]^{\frac{1}{2}}. \qquad (6.126)$$

The requirement that $|\psi(F_2)\rangle$ and $|\psi(F_1)\rangle$ be orthogonal gives (apart from an arbitrary phase factor)

$$|\psi(F_1)\rangle = b_J \left| {}^2\Pi_{\frac{3}{2}} vJ \right\rangle + a_J \left| {}^2\Pi_{\frac{1}{2}} vJ \right\rangle. \qquad (6.127)$$

E.

CONSTRUCT A PLOT OF $E({}^2\Pi; vJ)$ VERSUS Y FOR $B_v = 1$ CM^{-1}, $-20 \leq Y \leq 20$ AND $J = \frac{3}{2}, \frac{5}{2}$, AND $\frac{7}{2}$. OPTIONAL: INCLUDE THE LEVEL $J = \frac{1}{2}$ IN YOUR PLOT.

For $J = \frac{3}{2}$, $\frac{5}{2}$, and $\frac{7}{2}$, see Figure 6.4. For $J = \frac{1}{2}$, the energy is $E = \pm B_v \left(\frac{Y-2}{2} \right)$.

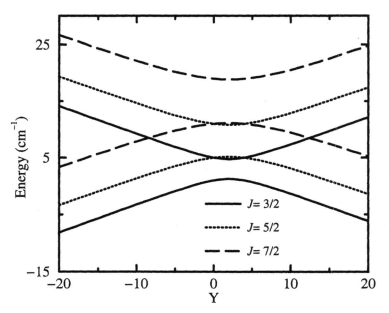

Figure 6.4: Plots of the energy $E(^2\Pi; vJ)$ for different values of J.

F.

THE ELECTRON CHARGE CLOUD ASYMMETRY IS GIVEN BY

$$\Delta = \langle \cos 2\chi \rangle, \tag{6.128}$$

WHERE χ IS THE AZIMUTHAL ANGLE OF THE ELECTRON CHARGE DISTRIBUTION IN THE MOLECULAR FRAME MEASURED FROM THE LINE OF NODES. SHOW THAT

$$\Delta = \pm a_J b_J$$

$$= \pm \left[\frac{X^2 - (Y-2)^2}{4X^2} \right]^{\frac{1}{2}} \tag{6.129}$$

BY EXPLICIT EVALUATION OF Δ USING THE F_1 AND F_2 WAVE FUNCTIONS. *Hint*: PRESENT ARGUMENTS THAT

$$\left\langle \psi \left(^2\Pi_{\frac{1}{2}} \right) \middle| \cos 2\chi \middle| \psi \left(^2\Pi_{\frac{1}{2}} \right) \right\rangle = \left\langle \psi \left(^2\Pi_{\frac{3}{2}} \right) \middle| \cos 2\chi \middle| \psi \left(^2\Pi_{\frac{3}{2}} \right) \right\rangle = 0 \tag{6.130}$$

WHEREAS

$$\left\langle \psi \left(^2\Pi_{\frac{1}{2}} \right) \middle| \cos 2\chi \middle| \psi \left(^2\Pi_{\frac{3}{2}} \right) \right\rangle = \left\langle \psi \left(^2\Pi_{\frac{3}{2}} \right) \middle| \cos 2\chi \middle| \psi \left(^2\Pi_{\frac{1}{2}} \right) \right\rangle = \frac{1}{2}. \tag{6.131}$$

The only part of the wave function that depends on χ is $|n\Lambda\rangle$. In calculating the expectation value of $\cos 2\chi$, we use $|S\Sigma\rangle$ to project out nonvanishing parts of the integral. It follows that

$$\left\langle \psi\left(^2\Pi_{\frac{1}{2}}\right)|\cos 2\chi|\,\psi\left(^2\Pi_{\frac{1}{2}}\right)\right\rangle = \frac{1}{2}\frac{1}{2\pi}\int_0^{2\pi} e^{-i\chi}\cos 2\chi\, e^{i\chi}\,d\chi + \frac{1}{2}\frac{1}{2\pi}\int_0^{2\pi} e^{i\chi}\cos 2\chi\, e^{-i\chi}\,d\chi$$

$$= \frac{1}{2\pi}\int_0^{2\pi}\cos 2\chi\,d\chi$$

$$= 0. \tag{6.132}$$

The same result is obtained for $\left\langle \psi\left(^2\Pi_{\frac{3}{2}}\right)|\cos 2\chi|\,\psi\left(^2\Pi_{\frac{3}{2}}\right)\right\rangle$. In the cross terms, however, there are nonvanishing contributions. For example,

$$\left\langle \psi\left(^2\Pi_{\frac{1}{2}}\right)|\cos 2\chi|\,\psi\left(^2\Pi_{\frac{3}{2}}\right)\right\rangle = \frac{1}{2}\frac{1}{2\pi}\int_0^{2\pi} e^{-i\chi}\cos 2\chi\, e^{-i\chi}\,d\chi + \frac{1}{2}\frac{1}{2\pi}\int_0^{2\pi} e^{i\chi}\cos 2\chi\, e^{i\chi}\,d\chi$$

$$= \frac{1}{2}\frac{1}{2\pi}\frac{1}{2}\int_0^{2\pi} e^{-i2\chi}\left(e^{i2\chi}+e^{-i2\chi}\right)d\chi$$

$$\quad + \frac{1}{2}\frac{1}{2\pi}\frac{1}{2}\int_0^{2\pi} e^{i2\chi}\left(e^{i2\chi}+e^{-i2\chi}\right)d\chi$$

$$= \frac{1}{2}. \tag{6.133}$$

We can now evaluate Δ as follows:

$$\Delta(F_1) = \left\langle \psi(F_1)|\cos 2\chi|\,\psi(F_1)\right\rangle$$

$$= -a_J b_J\left\langle ^2\Pi_{\frac{1}{2}};vJ\,|\cos 2\chi|^2\Pi_{\frac{3}{2}};vJ\right\rangle - b_J a_J\left\langle ^2\Pi_{\frac{3}{2}};vJ\,|\cos 2\chi|^2\Pi_{\frac{1}{2}};vJ\right\rangle$$

$$= -a_J b_J. \tag{6.134}$$

Similarly, for $\psi(F_2)$ we obtain

$$\Delta(F_2) = a_J b_J. \tag{6.135}$$

Multiplying equation 6.125 by equation 6.126 gives equation 6.129, as desired.

The above treatment, however, is only a simple approximation. In general, $|n\Lambda\rangle$ must be treated as $Y_{L\Lambda}(\theta, \chi)$ and the electron charge cloud asymmetry requires an integration over the full solid angle element. As pointed out by Y. Fujimura, this problem should be stated as follows: Show that

$$\Delta(F_1 p^{\pm}) = -\Delta(F_2 p^{\pm}) = \mp \left[\frac{X^2 - (Y-2)^2}{4X^2}\right]^{\frac{1}{2}} S \tag{6.136}$$

where

$$S = \left\langle J\Omega = -\frac{3}{2}M \middle| J\Omega = \frac{1}{2}M \right\rangle \tag{6.137}$$

is the overlap integral introduced by Green and Zare, *Chem. Phys.* **7**, 62 (1975), and $p^{\pm}$ denotes the parity block. In our $\chi = 0$ convention in which yz is the plane of rotation, S is close to unity for high J and high M. Specifically, for $M = J$, $S = \left[(J - \frac{1}{2}) / (J + \frac{3}{2})\right]^{\frac{1}{2}}$.

G.

FIND THE VALUES OF $|\Delta|$ VERSUS J FOR $\frac{1}{2} \le J \le \frac{9}{2}$ IN THE CASE OF THE TWO $^2\Pi$ GROUND STATE RADICALS, OH AND NO, GIVEN THAT $Y = -7.4$ FOR OH $X^2\Pi$ $(v = 0)$ AND $Y = 72.9$ FOR NO $X^2\Pi$ $(v = 0)$. THIS COMPARISON EXPLAINS WHY IT IS MORE LIKELY THAT PREFERENTIAL POPULATION OF THE Λ COMPONENTS WILL BE OBSERVED IN OH $X^2\Pi$ $(v = 0)$ THAN IN $X^2\Pi$ $(v = 0)$ FOR LOW TO MODERATE J VALUES IN A "COLLISION PROCESS."

Substituting the definition of X (equation 6.95) into equation 6.129, we obtain an expression for the J dependence of Δ,

$$|\Delta| = \left[\frac{(J + \frac{1}{2})^2 - 1}{4(J + \frac{1}{2})^2 + Y(Y - 4)}\right]^{\frac{1}{2}}. \tag{6.138}$$

Table 6.5: Electron charge cloud asymmetry parameter.

J	$|\Delta|$(OH)	$|\Delta|$ (NO)
1/2	0	0
3/2	0.172	0.024
5/2	0.258	0.039
7/2	0.318	0.054
9/2	0.361	0.068

Calculations using equation 6.138 and listed in Table 6.5 show that $|\Delta|$ ranges from 0 for $J = \frac{1}{2}$ to $\frac{1}{2}$ for $2J >> Y$. In the case of OH, $|\Delta(J = \frac{9}{2})| = 0.361$, whereas for NO, $|\Delta(J = \frac{9}{2})| = 0.068$.

H.

FOR A $^1\Pi$ STATE, THE TERMS IN THE HAMILTONIAN RESPONSIBLE FOR CAUSING INTERACTION BETWEEN THE $^1\Pi$ STATE AND OTHER $^1\Sigma^+$ AND $^1\Sigma^-$ STATES ARE

$$\mathcal{H}_1 = -B(r) \ [J_+L_- + J_-L_+] \tag{6.139}$$

(SEE EQUATION Z-5.15.10). USE A VAN VLECK TRANSFORMATION TO SHOW THAT FOR A v, J LEVEL OF PARITY p^+,

$$\mathcal{H}^{(2)}_{\Omega=1,\Omega=1} = q_v(^1\Sigma^+) \ J(J+1), \tag{6.140}$$

AND FOR A v, J LEVEL OF PARITY p^-,

$$\mathcal{H}^{(2)}_{\Omega=1,\Omega=1} = q_v(^1\Sigma^-) \ J(J+1), \tag{6.141}$$

WHERE

$$q_v \left(^1\Sigma^\pm\right) \ = \ \sum_{n'v'} \frac{\langle n \ ^1\Pi_1 v | B(r)L_+ | n' \ ^1\Sigma_0^\pm v' \rangle \langle n' \ ^1\Sigma_0^\pm v' | B(r)L_- | n \ ^1\Pi_1 v \rangle}{E\left(^1\Pi_1 v\right) - E\left(n' \ ^1\Sigma_0^\pm v'\right)}. \tag{6.142}$$

Because the parity of a state is unchanged by the perturbation operator, only $^1\Sigma^+$ states interact with p^+ levels of the $^1\Pi$ state, and only $^1\Sigma^-$ states interact with p^- levels of the $^1\Pi$ state. We use a second-order Van Vleck transformation (equation Z-6.104) to calculate the perturbation:

$$\mathcal{H}^{(2)} \ = \ \sum_{n'v'} \langle n \ ^1\Pi_1 v | \langle J1M | -B(r)J_-L_+ | n' \ ^1\Sigma_0^\pm v' \rangle | J0M \rangle$$

$$\times \ \frac{\langle n' \ ^1\Sigma_0^\pm v' | \langle J0M | -B(r)J_+L_- | n \ ^1\Pi_1 v \rangle | J1M \rangle}{E(n \ ^1\Pi_1 v) - E(n' \ ^1\Sigma_0^\pm v')}. \tag{6.143}$$

The matrix elements (equation Z-3.49) of $J_\pm$ are

$$\langle J0M | J_+ | J1M \rangle \ = \ \langle J1M | J_- | J0M \rangle$$

$$= \ \sqrt{J(J+1)}. \tag{6.144}$$

Inserting equation 6.144 into equation 6.143, we obtain

$$\mathcal{H}^{(2)} = q_v(^1\Sigma^\pm)J(J+1), \tag{6.145}$$

with q_v as defined in equation 6.142.

I.

For a Q branch member of a $^2\Sigma - {}^2\Sigma$ transition one of the wave functions will be

$$\frac{1}{\sqrt{2}}\left[\left|n0\right\rangle\left|\frac{1}{2}\frac{1}{2}\right\rangle\left|J\frac{1}{2}M\right\rangle + \left|n0\right\rangle\left|\frac{1}{2}-\frac{1}{2}\right\rangle\left|J-\frac{1}{2}M\right\rangle\right] \tag{6.146}$$

and the other will be

$$\frac{1}{\sqrt{2}}\left[\left|n0\right\rangle\left|\frac{1}{2}\frac{1}{2}\right\rangle\left|J\frac{1}{2}M\right\rangle - \left|n0\right\rangle\left|\frac{1}{2}-\frac{1}{2}\right\rangle\left|J-\frac{1}{2}M\right\rangle\right] \tag{6.147}$$

so that these two wavefunctions have the same value of J but are of opposite parity. The rotational line strength is calculated using equation Z-6.15.66. Specifically,

$$
\begin{aligned}
S(J, J) &= (2J+1)^2\left[\frac{1}{2}(-1)^{J-1+\frac{1}{2}}\begin{pmatrix} J & 1 & J \\ \frac{1}{2} & 0 & -\frac{1}{2} \end{pmatrix} - \frac{1}{2}(-1)^{J-1-\frac{1}{2}}\begin{pmatrix} J & 1 & J \\ -\frac{1}{2} & 0 & \frac{1}{2} \end{pmatrix}\right]^2 \\[2ex]
&= (2J+1)^2\left[\frac{1}{2}(-1)^{J-1+\frac{1}{2}}\begin{pmatrix} J & 1 & J \\ \frac{1}{2} & 0 & -\frac{1}{2} \end{pmatrix} - \frac{1}{2}(-1)^{J-1-\frac{1}{2}}(-1)^{2J+1}\begin{pmatrix} J & 1 & J \\ \frac{1}{2} & 0 & -\frac{1}{2} \end{pmatrix}\right]^2 \\[2ex]
&= (2J+1)^2\begin{pmatrix} J & 1 & J \\ \frac{1}{2} & 0 & -\frac{1}{2} \end{pmatrix}^2 \\[2ex]
&= (2J+1)^2\frac{1}{(2J+2)(2J+1)(2J)} \\[2ex]
&= \frac{2J+1}{4J(J+1)}. \tag{6.148}
\end{aligned}
$$

APPLICATION 16
MOLECULAR REORIENTATION IN LIQUIDS

A.

SHOW THAT

$$\frac{\partial P(\Omega, t)}{\partial t} = -\mathbf{L} \cdot \underset{\sim}{\mathbf{D}} \cdot \mathbf{L} \, P(\Omega, t),$$

(6.149)

WHERE

$$\underset{\sim}{\mathbf{D}} = \frac{1}{2\,\Delta t} \int \boldsymbol{\theta}\boldsymbol{\theta} p(\theta, \Delta t) \mathrm{d}^3\theta$$

(6.150)

IS THE SECOND MOMENT, $\frac{1}{2} \langle \boldsymbol{\theta}\boldsymbol{\theta} \rangle / \Delta t$, OR THE SO-CALLED DIFFUSION TENSOR. HERE $\langle \cdots \rangle$ DENOTES AN ENSEMBLE AVERAGE. *Hint:* SUBSTITUTE EQUATIONS Z-6.16.2 AND Z-6.16.3 INTO EQUATION Z-6.16.1, EXPAND THE EXPONENTIAL IN THE INTEGRAND, DEVELOP $P(\Omega, t + \Delta t)$ IN A TAYLOR SERIES IN Δt, AND RETAIN ONLY TERMS FIRST ORDER IN Δt AND SECOND ORDER IN θ. USE THE FACT THAT THE FIRST MOMENT,

$$\frac{\langle \boldsymbol{\theta} \rangle}{\Delta t} = \frac{1}{\Delta t} \int \boldsymbol{\theta} p(\theta, t) \, \mathrm{d}^3\theta,$$

(6.151)

VANISHES FOR A RANDOM WALK.

From equation Z-6.16.3, we know that

$$P(\Omega_0, t)\mathrm{d}\Omega_0 = e^{i\boldsymbol{\theta} \cdot \mathbf{L}} \, P(\Omega, t)\mathrm{d}\Omega,$$

(6.152)

where $\boldsymbol{\theta}$ is the angle conjugate to $\mathbf{L}$ which brings Ω_0 to Ω. ($\boldsymbol{\theta} = \Delta\Omega\,\hat{n}$, with $\hat{n}$ denoting the unit vector along the axis of rotation.) Inserting equation 6.152 into Z-6.16.1, we obtain

$$P(\Omega, t + \Delta t)\mathrm{d}\Omega = \int p(\Delta\Omega, \Delta t) \, \mathrm{d}(\Delta\Omega) \, e^{i\boldsymbol{\theta} \cdot \mathbf{L}} \, P(\Omega, t)\mathrm{d}\Omega.$$

(6.153)

The Taylor series expansion of $P(\Omega, t + \Delta t)$ in Δt yields

$$P(\Omega, t + \Delta t) = P(\Omega, t) + \frac{\partial P(\Omega, t)}{\partial t} \Delta t + \frac{1}{2!} \frac{\partial^2 P(\Omega, t)}{\partial t^2} (\Delta t)^2 + \ldots$$

(6.154)

We can also expand $e^{i\boldsymbol{\theta} \cdot \mathbf{L}}$ in a Taylor series in $\boldsymbol{\theta} \cdot \mathbf{L}$ because $\Delta\Omega$ is small when Δt is small:

$$e^{i\boldsymbol{\theta} \cdot \mathbf{L}} = 1 + i\boldsymbol{\theta} \cdot \mathbf{L} - \frac{1}{2}(\boldsymbol{\theta} \cdot \mathbf{L})(\boldsymbol{\theta} \cdot \mathbf{L}) + \ldots$$

(6.155)

Inserting equations 6.154 and 6.155 into equation 6.153, and dropping $\mathrm{d}\Omega$, we obtain

$$P(\Omega, t) + \frac{\partial P(\Omega, t)}{\partial t} \Delta t + \frac{1}{2!} \frac{\partial^2 P(\Omega, t)}{\partial t^2} (\Delta t)^2 + \ldots$$

$$= \int p(\Delta\Omega, \Delta t) \, \mathrm{d}(\Delta\Omega)P(\Omega, t) + i\int p(\Delta\Omega, \Delta t)(\boldsymbol{\theta} \cdot \mathbf{L})\mathrm{d}(\Delta\Omega)P(\Omega, t)$$

$$- \frac{1}{2}\int p(\Delta\Omega, \Delta t)(\boldsymbol{\theta} \cdot \mathbf{L})(\boldsymbol{\theta} \cdot \mathbf{L})\mathrm{d}(\Delta\Omega)P(\Omega, t) + \ldots$$

(6.156)

From the definition of a probability, we know that

$$\int p(\Delta\Omega, \Delta t) \, \mathrm{d}(\Delta\Omega) = 1. \tag{6.157}$$

Because the molecular motion is assumed to be a random walk over small angular orientation, we use the fact that $p(\Delta\Omega, \Delta t)$ is an even function of $\Delta\Omega$. Therefore,

$$\int p(\Delta\Omega, \Delta t)(\boldsymbol{\theta} \cdot \mathbf{L})\mathrm{d}(\Delta\Omega) = \int p(\Delta\Omega, \Delta t) \, \Delta\Omega \, \mathrm{d}(\Delta\Omega)\hat{n} \cdot \mathbf{L}$$

$$= 0. \tag{6.158}$$

Inserting equations 6.157 and 6.158 into equation 6.156, we get

$$\frac{\partial P(\Omega, t)}{\partial t} \Delta t + \frac{1}{2!} \frac{\partial^2 P(\Omega, t)}{\partial t^2} (\Delta t)^2 + \dots = -\frac{1}{2} \int p(\Delta\Omega, \Delta t)(\boldsymbol{\theta} \cdot \mathbf{L})(\boldsymbol{\theta} \cdot \mathbf{L})\mathrm{d}(\Delta\Omega)P(\Omega, t) + \dots. \tag{6.159}$$

We are interested in developing the differential relation where $\Delta t \to 0$ and $\Delta\Omega \to 0$. We can take from each side only the first nonvanishing member of the Taylor series, obtaining

$$\frac{\partial P(\Omega, t)}{\partial t} \Delta t = -\frac{1}{2} \int p(\Delta\Omega, \Delta t)(\boldsymbol{\theta} \cdot \mathbf{L})(\boldsymbol{\theta} \cdot \mathbf{L})\mathrm{d}(\Delta\Omega)P(\Omega, t)$$

$$= -\frac{1}{2}\mathbf{L} \cdot \int p(\Delta\Omega, \Delta t)\boldsymbol{\theta}\boldsymbol{\theta}\mathrm{d}(\Delta\Omega) \cdot \mathbf{L}P(\Omega, t). \tag{6.160}$$

Finally, using the definition of the diffusion tensor given in equation 6.150, we obtain

$$\frac{\partial P}{\partial t}(\Omega, t) = -\mathbf{L} \cdot \underset{\sim}{\mathbf{D}} \cdot \mathbf{L}P(\Omega, t). \tag{6.161}$$

B.

THE GREEN'S FUNCTION $G(\Omega_0 | \Omega, t)$ IS THE CONDITIONAL PROBABILITY DENSITY THAT IF THE DIFFUSOR WAS INITIALLY AT Ω_0 THEN AT TIME t IT WILL BE FOUND AT Ω. IT DESCRIBES THE EVOLUTION OF THE DIFFUSOR FROM ITS ORIGINAL ORIENTATION, SUBJECT TO THE INITIAL CONDITION THAT

$$G(\Omega_0 | \Omega, 0) = \delta^3(\Omega - \Omega_0). \tag{6.162}$$

BECAUSE $P(\Omega, t)$ SATISFIES THE DIFFUSION EQUATION, SO MUST $G(\Omega_0 | \Omega, t)$; THAT IS,

$$\frac{\partial G(\Omega_0 | \Omega, 0)}{\partial t} = -H \, G(\Omega_0 | \Omega, 0). \tag{6.163}$$

SHOW THAT

$$G(\Omega_0 | \Omega, t) = \sum_n \psi_n^*(\Omega_0) \, \psi_n(\Omega_0) \, e^{-E_n t}. \tag{6.164}$$

USE THIS RESULT TO EXPRESS THE GREEN'S FUNCTION FOR A SYMMETRIC DIFFUSOR AS

$$G(\Omega_0|\Omega,t) \;=\; \sum_{L=0}^{\infty} \sum_{M,K=-L}^{L} \frac{2L+1}{8\pi^2} D_{MK}^{L}(\phi_0,\theta_0,\chi_0)\, D_{MK}^{L*}(\phi,\theta,\chi) e^{-[D_\perp L(L+1)+(D_\parallel - D_\perp)K^2]t}.$$

$$(6.165)$$

Let $\psi_n(\Omega)$ be the normalized eigenfunction of H with eigenvalues E_n. Because $\{\psi_n(\Omega)\}$ forms a complete set, we can use it to expand $G(\Omega_0|\Omega,t)$:

$$G(\Omega_0|\Omega,t) = \sum_n A_n(t)\psi_n(\Omega), \qquad (6.166)$$

where

$$A_n(t) = \int d\Omega\ \psi_n^*(\Omega)\ G(\Omega_0|\Omega,t). \qquad (6.167)$$

From equation 6.162, we obtain for $A_n(0)$

$$A_n(0) = \int d\Omega\ \psi_n^*(\Omega)\ G(\Omega_0|\Omega,0)$$

$$= \psi_n^*(\Omega_0). \qquad (6.168)$$

Inserting equation 6.166 into equation 6.163, we get

$$\frac{\partial \sum_n A_n(t)\psi_n}{\partial t} = -H\sum_n A_n(t)\psi_n$$

$$= -\sum_n E_n A_n(t)\psi_n. \qquad (6.169)$$

Because $\{\psi_n(\Omega)\}$ forms an orthonormal basis set, equation 6.169 implies that

$$\frac{\partial A_n(t)}{\partial t} = -E_n A_n(t), \qquad (6.170)$$

which can be solved to obtain

$$A_n(t) = A_n(0)\, e^{-E_n t}. \qquad (6.171)$$

Inserting equations 6.168 and 6.171 into equation 6.166 gives

$$G(\Omega_0|\Omega,t) = \sum_n \psi_n^*(\Omega_0)\ \psi_n(\Omega)\ e^{-E_n t}. \qquad (6.172)$$

For a symmetric diffusor, $\psi_n(\Omega)$ is given by equation Z-3.125,

$$\psi_n(\Omega) = \left[\frac{2L+1}{8\pi^2}\right]^{\frac{1}{2}} D_{MK}^{L*}(\phi,\theta,\chi), \qquad (6.173)$$

and E_n is given by equation Z-6.51 with A and C replaced by $D_\perp$ and $D_\parallel$, respectively,

$$E_n = D_\perp L(L+1) + (D_\parallel - D_\perp)K^2. \tag{6.174}$$

Note that in equations 6.173 and 6.174 n represents collectively L, K, M. Inserting these two equations into equation 6.172, we obtain equation 6.165, as desired.

C.

SHOW THAT FOR A SYMMETRIC DIFFUSOR WITH AN ISOTROPIC INITIAL DISTRIBUTION,

$$F_{MK}^L(t) = \frac{1}{8\pi^2} e^{-[D_\perp L(L+1) + (D_\parallel - D_\perp)K^2]t} \tag{6.175}$$

For a symmetric diffusor, $F_{MK}^L(t)$ is given by

$$F_{MK}^L(t) = \int d\Omega_0 \int d\Omega \, [G(\Omega_0|\Omega, t)P(\Omega_0)\Psi_{LKM}(\Omega_0)]^* \psi_{LKM}(\Omega), \tag{6.176}$$

where the expressions for $G(\Omega_0|\Omega, t)$ and Ψ_{LKM} are given by equations 6.165 and 6.173, respectively. For an isotropic initial distribution,

$$P(\Omega_0) = \frac{1}{8\pi^2}, \tag{6.177}$$

where $P(\Omega_0)$ satisfies the normalization

$$\int P(\Omega_0) \, d\Omega_0 = 1. \tag{6.178}$$

Inserting equations 6.165, 6.173, and 6.178 into equation 6.176, we obtain

$$F_{MK}^L(t) = \sum_{M'K'L'} \int d\Omega_0 \int d\Omega \, \frac{2L'+1}{8\pi^2} D_{M'K'}^{L'*}(\phi_0, \theta_0, \chi_0) D_{M'K'}^{L'}(\phi, \theta, \chi)$$

$$\times \, e^{-[D_\perp L'(L'+1) + (D_\parallel - D_\perp)K'^2]t} \frac{1}{8\pi^2} \frac{2L+1}{8\pi^2} D_{MK}^L(\phi_0, \theta_0, \chi_0) \, D_{MK}^{L*}(\phi, \theta, \chi). \tag{6.179}$$

Equation 6.179 can be simplified with the help of equation Z-3.113,

$$F_{MK}^L(t) = \frac{1}{8\pi^2} \sum_{L'M'K'} \delta_{LL'} \, \delta_{MM'} \, \delta_{KK'} e^{-[D_\perp L'(L'+1) + (D_\parallel - D_\perp)K'^2]t}$$

$$= \frac{1}{8\pi^2} e^{-[D_\perp L(L+1) + (D_\parallel - D_\perp)K^2]t}. \tag{6.180}$$

Chapter 7

ERRATA TO 2ND PRINTING OF *ANGULAR MOMENTUM*

The following errata have been collected from many sources to which we express our sincere gratitude. Special thanks go to Melissa A. Hines, Alexei Buchachenko, Shan Tao Lai, Andrew J. Orr-Ewing, and Yo Fujimura who communicated numerous corrections.

pg. 10 In Eq. (1.57) for the expression for $Y_{2,\pm 1}(\theta, \phi)$ change 16π to 8π in the square root sign.

pg. 36 In Eq. (45) change $-$ sign to $+$ sign in front of $\frac{1}{b}(\frac{\pi}{2})$.

pg. 36 In Eq. (46) replace $f[b(x), y]$ by $f[x, b(x)]$ and replace $f[a(x), y]$ by $f[x, a(x)]$.

pg. 36 In the two lines below Eq. (46) replace b/k by $1/k$.

pg. 38 Replace twice k by κ in the expression in the line below Eq. (58).

pg. 39 In the ordinate of Fig. 5 replace $E_0 - E$ by $(E_0 - E)$.

pg. 43 In the third line below Eq. (2.1) replace Eq. (1.16) by Eq. (1.8).

pg. 66 In line 9 replace "total angular momentum" by "total orbital angular momentum."

pg. 90 In both Eq. (3.77) and Eq. (3.78) replace superscript $\dagger$ by superscript $*$, and replace $d_{MM'}^{J}(\theta)$ by $d_{M'M}^{J}(\theta)$.

pg. 91 In line 1 replace $\dagger$ by $*$ and omit "transpose."

pg. 99 Fig. 3.6 is misdrawn; $\mathbf{J}_2$ should connect to $\mathbf{J}_1$ and $\mathbf{J}_3$.

pg. 99 In the second to last line on the page change "Figures 1.1, 1.2, 1.3, and 2.2" to read "Figures 1.1, 2.1, 2.2, and 2.3."

pg. 101 In the exponent of (-1) in Eq. (3.111) change $J_1 - M_1$ to $J_1 + M_1$.

pg. 101 In the exponent of (-1) in Eq. (3.112) change $J_1 - M_1'$ to $J_1 + M_1'$.

pg. 117 In the left side of Eq. (2) replace $P_{JM}(\theta)$ by $P_{JM}(\theta)\mathrm{d}\Omega$.

pg. 144 In Eq. (4.3) replace $\langle j_{12}\, j_3\, j \,|\, j_1\, j_{23}\, j' \rangle$ by $\langle j_{12}\, j_3\, j' \,|\, j_1\, j_{23}\, j \rangle$.

pg. 149 In the second row of the 9-j symbol appearing once in Eq. (4.21) and twice in Eq. (4.22) interchange j_3 and j_4.

pg. 152 Rewrite rules 1 and 2 to read:

1. To add an arrow pointing toward or to drop an arrow pointing away from a particular jm pair, multiply the diagram by $(-1)^{j+m}$ and change the sign of m in the diagram.

2. To add an arrow pointing away or drop an arrow pointing toward a particular jm pair, multiply the diagram by $(-1)^{j-m}$ and change the sign of m in the diagram.

pg. 164 In the graphical diagram at the bottom of the page omit the central line $j_{1234} = 0$ and its label and add two lines connecting the $+$ and $-$ nodes; label these lines by j_{12} and j_{34}.

pg. 167 In Eq. (4.58) in the graph in the first line at the top of the page change from $+$ to $-$ the node on the right and the node on the bottom. Then in the graph in the second line of this equation, change from $+$ to $-$ the node on the top and the central node. Finally, remove the phrase starting with "where . . ."

pg. 174 In Eq. (4.65) change $\mathbf{j_2}$ to $\mathbf{j}$.

pg. 174 In the first line below Eq. (4.67) change $\mathbf{j_{23}}$ to $|\mathbf{j_{23}}|$.

pg. 174 In Eq. (4.67) add a minus sign before

$$\frac{d}{dt}(\mathbf{j_{23}} \cdot \mathbf{j}).$$

pg. 174 In Eq. (4.69) change $(\mathbf{j_1} \times \mathbf{j})$ to $(\mathbf{j_{12}} \times \mathbf{j})$.

pg. 176 In Eq. (4.71) change j_9 to j_7 in the last 6-j symbol on the right side of this equation.

pg. 178 In the third line from the bottom, change (see Application 4) to (see Eq. (14) of Application 4).

pg. 179 In three lines below Eq. (5.8) change Eq. (5.9) to Eq. (5.8).

pg. 181 In the line below Eq. (5.14) insert the factor $(-1)^{k+j-j'}$ in front of $(2j + 1)^{1/2}$.

pg. 183 In the second line of Eq. (5.25) replace $\langle 3 \| L \| 3 \rangle$ by $\langle 3 \| L^{(1)} \| 3 \rangle$.

pg. 187 In Eq. (5.38) change $\sum_{q'}$ to $\sum_{q',q_1'}$.

pg. 188 In Eq. (5.46) in the first and second lines replace $\sum_q$ by $\sum_{q,q'}$.

pg. 189 In the line above Eq. (5.50) replace Eq. (4.21) by Eq. (4.20).

pg. 190 In the last line of the first paragraph replace "the gradient of the electric field" by "the gradient of the gradient of the electric field."

pg. 194 In Eq. (5.67) add $+2j_1$ to the exponent of (-1).

pg. 194 In Eq. (5.69) add $\delta_{jj'}$ to the last line on the right side.

pg. 195 In Eq. (5.70) add on the right side $\delta_{jj'}$.

pg. 201 In Eq. (5.85) replace $\psi(\alpha_e J_e M_e; t = 0$ by $\psi(\alpha_e J_e; t = 0)$.

pg. 201 In Eqs. (5.86), (5.87), and (5.88) replace $\psi(\alpha_e J_e M_e; t)$ by $\psi(\alpha_e J_e; t)$.

pg. 204 In the last line of this page remove the parentheses about $\hat{\mathbf{e}}_a$ and $\hat{\mathbf{e}}_a^*$.

pg. 209 In Eq. (5.118) insert a minus sign before the right side.

pg. 209 In two lines below Eq. (5.118) insert a minus sign before $e(1, -1)$.

pg. 216 In Eq. (5.133) replace † by *.

pg. 217 In Eq. (5.134) and in Eq. (5.136) change † to * twice in each equation.

pg. 222 In Eq. (13) second line from bottom of page change j' to j.

pg. 224 In right side of Eq. (17) change -1 to 1.

pg. 228 Below Eq. (23) it should read: In the general case our system is in a *mixed state* which is represented by a density operator ρ that is an incoherent superposition of a number of pure states $|\psi^{(i)}\rangle$ with statistical weights $W^{(i)}$.

$$\rho = \sum_i W^{(i)} \left|\psi^{(i)}\right\rangle \left\langle\psi^{(i)}\right| \tag{24}$$

where

$$\sum_i W^{(i)} = 1. \tag{25}$$

pg. 228 In the line below Eq. (25) change "in" to "In."

pg. 229 In Eq. (27) insert "$\langle JM|\rho|JM'\rangle =$" before the expression on the right side of this equation.

pg. 229 In the sixth line from the bottom of the page, change a_M to $a_M^{(i)}$.

pg. 230 Change the sentence in line 13 to read: A pure state represents a completely ordered ensemble, whereas the mixed state that is uniform is in a state of maximum disorder.

pg. 230 Change the sentence in line 17 to read: Then for a pure state $S = 0$, whereas for a state of maximum disorder $S = \ln(2J + 1)$.

pg. 232 Add "cosine of the" before "angle" in part J.

pg. 237 In the exponent of (-1) in Eqs. (62) and (63) change M' to M.

pg. 238 In Eqs. (67) and (69) insert 3 in front of $O_0^{(1)}(J_i)$ and after $A_0^{(2)}$ insert the factor

$$\left[\frac{5J_i(J_i + 1)}{(2J_i + 3)(2J_i - 1)}\right].$$

pg. 238 In Eq. (68) change = to $\propto$.

pg. 239 In the right side of Eq. (71) change $A_0^{(2)}(J_f)$ to $A_0^{(2)}(J_i)$.

pg. 239 In the right side of Eq. (72) it should read:

$$\left[1 - \frac{1}{(J_f + 1)^2}\right]^{1/2}.$$

pg. 239 In the right side of Eq. (73) change $(2J_f + 3)$ to $(2J_f + 1)$.

pg. 239 In the right side of Eq. (76) make it read: $\left[1 - \frac{1}{J_f^2}\right]^{1/2}$.

pg. 239 In the right side of Eq. (77) change $(2J_f - 1)$ to $(2J_f + 1)$.

pg. 240 In the Eqs. (79) and (80) drop the subscript q on J and the subscript 0 on I.

pg. 240 In the right side of Eq. (81) insert the factor $[(2F' + 1)(2F + 1)]^{1/2}$ inside the summation sign.

pg. 241 In Eq. (84) add a minus sign in front of the factor $i(E_{F'} - E_F)t/\hbar$ that appears in the first and third lines of this equation.

pg. 250 In the fourth line of Eq. (22) add a minus sign in front of $(5)^{1/2}$.

pg. 271 Replace the paragraph starting with "Eq. (6.63) is ..." by "Note that Eq. (6.58), (6.59), and (6.60) are valid for integer and half-integer J and K, whereas Eq. (6.63) holds true only for integral J and K."

pg. 280 In Eq. (6.99) insert i in front of $(E_j - E_k)$.

pg. 281 In line 2 change Eq. (6.97) to Eq. (6.98).

pg. 281 In Eq. (6.103) in the first line for $\langle j| \mathcal{G}_3 |k\rangle$ change $(E_k - E_\alpha)$ to $(E_k - E_\beta)$.

pg. 282 In Eq. (6.105) in its first line, replace twice $\mathcal{H}_R$ by $\mathcal{H}_1$.

pg. 284 In Eq. (6.112) replace π by π^4.

pg. 285 Eq. (6.117) should read:

$$
\begin{aligned}
S(J'K'; J''K'') \\
&= 3 \sum_{M',M''} \left| \left[\left(\frac{2J'+1}{8\pi^2}\right) \left(\frac{2J''+1}{8\pi^2}\right) \right]^{1/2} \int D^{J'}_{M'K'} D^{1*}_{0K'-K''} D^{J''*}_{M''K''} d\Omega \right|^2 \\
&= 3 \sum_{M',M''} \left| \frac{[(2J'+1)(2J''+1)]^{1/2}}{8\pi^2} \left[\int D^{J'*}_{M'K'} D^{1}_{0K'-K''} D^{J''}_{M''K''} d\Omega \right]^* \right|^2 \\
&= 3 \sum_{M',M''} \left| \left(\frac{2J''+1}{2J'+1}\right) \langle J''M'', 10|J'M'\rangle \langle J''K'', 1K'-K''|J'M'\rangle \right|^2 \\
&= 3 \frac{2J''+1}{2J'+1} \langle J''K'', 1K'-K''|J'K'\rangle^2 \sum_{M',M''} \langle J''M'', 10|J'M'\rangle^2 \\
&= 3 \frac{2J''+1}{2J'+1} \langle J''K'', 1K'-K''|J'K'\rangle^2 \sum_{M} \left(\frac{2J'+1}{3}\right) \langle J''M'', J'-M|10\rangle^2 \\
&= (2J''+1) \langle J''K'', 1K'-K'|J'K'\rangle^2 \\
&= (2J'+1)(2J''+1) \begin{pmatrix} J'' & 1 & J' \\ K'' & K'-K'' & -K' \end{pmatrix}^2
\end{aligned}
$$

pg. 287 In Eq. (6.123) insert the phase factor $(-1)^{J'-1+K''}$ in front of the 3-j symbol inside the double summation.

pg. 294 In Eq. (6.142) change $\boldsymbol{\omega}$ to ω in the middle term.

pg. 297 In last line insert after rotation: for $M = J$.

pg. 298 In line 2 insert after rotation: for $M = J$.

pg. 303 Change Eq. (34) to read:

$$|\psi(F_2)\rangle = a_J \left|{}^2\Pi_{\frac{3}{2}} vJ\right\rangle - b_J \left|{}^2\Pi_{\frac{1}{2}} vJ\right\rangle.$$

pg. 303 Change Eq. (35) to read:

$$|\psi(F_1)\rangle = b_J \left|{}^2\Pi_{\frac{3}{2}} vJ\right\rangle + a_J \left|{}^2\Pi_{\frac{1}{2}} vJ\right\rangle.$$

pg. 304 In the second sentence change F_1 and F_2 to F_2 and F_1.

pg. 306 In line 13 change former to latter.

pg. 306 Change Eq. (38) to read:

$$\sigma_v(yz)\, |n \Lambda\rangle = (-1)^s \, |n -\Lambda\rangle.$$

pg. 306 Change Eq. (39) to read:

$$\sigma_v(yz)\, |S \Sigma\rangle = (-1)^s \, |S -\Sigma\rangle.$$

pg. 306 In the first line above Eq. (40) change xz to yz and change $\chi \to -\chi$ to $\chi \to \pi - \chi$.

pg. 306 Change Eq. (40) to read:

$$Y_{L\Lambda}(\theta, \chi) \to Y_{L\Lambda}(\theta, \pi - \chi) = (-1)^\Lambda Y_{L\Lambda}^*(\theta, \chi) = Y_{L-\Lambda}(\theta, \chi).$$

pg. 307 In line 3 change $\sigma_v(xz), y \to -y$ to $\sigma_v(yz), x \to -x$.

pg. 307 In line 5 change y to x and change $C_2(y)$ to $C_2(x)$.

pg. 307 In line 6 change $C_2(y)$ to $C_2(x)$.

pg. 307 Change Eqs. (41) and (42) by replacing $(0, \pi, 0)$ by $(\pi, \pi, 0)$; the right side of Eq. (41) is $e^{i\pi/2} \left|\frac{1}{2}, -\frac{1}{2}\right\rangle$; and the right side of Eq. (42) is $e^{i\pi/2} \left|\frac{1}{2}, \frac{1}{2}\right\rangle$.

pg. 307 Change Eq. (43) to read:

$$\sigma_v(yz) \left|\frac{1}{2}, \sigma\right\rangle = (-1)^{1/2} \left|\frac{1}{2}, -\sigma\right\rangle.$$

pg. 307 In line 9 of the second paragraph the sentence should read: On reflection we interchange α and β in the uncoupled state so that $\Sigma \to -\Sigma$, and we pick up a phase factor $(-1)^x$, where x equals the number of electrons divided by two, i.e., $x = n/2$. Then omit the sentence beginning with "Since..."

pg. 307 In Eq. (45) change $\sigma_v(xz)$ to $\sigma_v(yz)$, and change $(-1)^{S-\Sigma}$ to $(-1)^S$.

pg. 308 In the third line below Eq. (46) change $x \to x$, $y \to -y$, and $z \to z$ to read: $x \to -x$, $y \to y$, and $z \to z$.

pg. 308 In the fourth line below Eq. (46) change $\pi - \chi$ to $-\chi$ and change $C_2(y)$ to $C_2(x)$.

pg. 308 In the fifth line below Eq. (46) change Eq. (6.59) to Eq. (6.60).

pg. 308 In Eq. (47) change $J - \Omega$ to $-J$.

pg. 308 In Eq. (48) in the second line change $\Lambda + s$ to s, $S - \Sigma$ to S, and $J - \Omega$ to $-J$.

pg. 312 Replace $a_{n\Omega}$ by $a_{n\Sigma\Omega}$ in Eq. (65) and twice in the second line from the bottom.

pg. 312 In the third line from the bottom of the page, change $\delta_{0\Lambda}$ to $\delta_{0,\Lambda}$.

pg. 313 Change Eq. (66) to read:

$$
S(J'; J) = (2J' + 1)(2J + 1)
$$
$$
\times \left| \sum_{\Omega'} \sum_{\Omega} a_{n'\Sigma'\Omega'}(p'^{\pm}) a_{n\Sigma\Omega}(p^{\pm}) \delta_{\Sigma,\Sigma'} (-1)^{J'-1+\Omega} \begin{pmatrix} J & 1 & J' \\ \Omega & \Omega' - \Omega & -\Omega' \end{pmatrix} \right|^2 .
$$

pg. 316 In reference 27, change $C_2(x)\sigma_v(yz)$ to $C_2(y)\sigma_v(xz)$.

pg. 316 Update the last line of reference 29 to read: M. H. Alexander et al. *J. Chem. Phys.* **89**, 1749 (1988).

pg. 317 Reference 31 should be rewritten to read:

As Alexander and Dagdigian [29] show, the electron distribution in the F_1 e Λ-doublet level is oriented preferentially in the plane of rotation and in the F_1 f Λ-doublet level oriented preferentially perpendicular to the plane of rotation along **J** for a single filled π orbital. The opposite applies to F_2 Λ-doublet levels. For a π^3 configuration, the preferences are reversed. The treatment outlined in F is an approximation in which the integration over only χ is considered. A more complete treatment in which the integration over θ and ϕ are included modifies Eq. (50).

pg. 320 In Eq. 16 replace $\psi_{LKM}(\Omega_0)$ by $\psi_{LKM}(\Omega)$ and replace $\psi_{LKM}(\Omega)$ by $\psi_{LKM}(\Omega_0)$.

pg. 325 Eq. (A-3) should have the signs in front of ν in the parentheses in the denominator $-, -, -, +, +$ so that it agrees with Eq. (2.25).